T0326633

Rural Poverty Determinants in the Remote Rural Areas of Kyrgyzstan

Schriften zur Internationalen Entwicklungs- und Umweltforschung

Herausgegeben vom

Zentrum für internationale Entwicklungs- und Umweltforschung

der Justus-Liebig-Universität Gießen

Band 33

Kanat Tilekeyev

Rural Poverty Determinants in the Remote Rural Areas of Kyrgyzstan

A Production Efficiency Impact on
the Poverty Level of a Rural Household

Bibliographic Information published by the Deutsche Nationalbibliothek
The Deutsche Nationalbibliothek lists this publication in the Deutsche Nationalbibliografie; detailed bibliographic data is available in the internet at http://dnb.d-nb.de.

Zugl:: Gießen, Fachbereich Wirtschaftswissenschaften, Univ., Diss., 2012

Gedruckt mit Unterstützung des
Deutschen Akademischen Austauschdienstes
und der University of Central Asia.

Printed with financial support of the German Academic Exchange Service
and the University of Central Asia.

Library of Congress Cataloging-in-Publication Data
Tilekeyev, Kanat, 1972-
Rural poverty determinants in the remote rural areas of Kyrgyzstan : a production efficiency impact on the poverty level of a rural household / Kanat Tilekeyev.
    pages cm
Includes bibliographical references.
ISBN 978-3-631-65093-6 -- ISBN 978-3-653-04289-4 (ebook) 1. Rural poor--Kyrgyzstan. 2. Poverty--Kyrgyzstan. 3. Rural development--Kyrgyzstan. 4. Agriculture--Economic aspects--Kyrgyzstan. I. Title.
HC420.7.Z9P627 2014
339.4'6095843--dc23
                        2014014948

ISSN 1615-312X
ISBN 978-3-631-65093-6 (Print)
E-ISBN 978-3-653-04289-4 (E-Book)
DOI 10.3726/ 978-3-653-04289-4

© Peter Lang GmbH
International Academic Publishers
Frankfurt am Main 2014
All rights reserved.
PL Academic Research is an Imprint of Peter Lang GmbH

TABLE OF CONTENTS

LIST OF FIGURES

LIST OF TABLES

LIST OF ABBREVIATIONS

AE	-	Allocative Efficiency	NGO	-	Non-Governmental Organization
CD	-	Cobb-Douglas production function	NDRS	-	Non-Decreasing Returns to Scale
COLS	-	Corrected Ordinary Least Squares	NIRS	-	Non-Increasing Returns to Scale
CPI	-	Consumer Price Index			
CRS	-	Constant Returns to Scale	NSC	-	National Statistical Committee
DAAD	-	Deutscher Akademischer Austausch Dienst (German Academic Exchange Service)	OLS	-	Ordinary Least Squares
DEA	-	Data Envelopment Analysis	PDF	-	Probability Density Function
DFID	-	UK Department for International Development	PDI	-	Poverty Distance Index
FAO	-	Food and Agriculture Organization	PG	-	Poverty Gap
FGT	-	Foster-Greer-Thorbecke Index	PP	-	Population Points
GDP	-	Gross Domestic Product	PPP	-	Purchasing Power Parity
HBS	-	Household Budget Survey	PPS	-	Probability Proportional to Size
HCI	-	Headcount Index			
HES	-	Household Energy Survey	PSI	-	Poverty Severity Index
HH	-	Household	PSU	-	Primary Sampling Unit
IV	-	Instrumental Variables	SFA	-	Stochastic Frontier Analysis
KIHS	-	Kyrgyz Integrated Household Survey			
KMPS	-	Kyrgyz Multipurpose Poverty Survey	TE	-	Technical Efficiency
KPMS	-	Kyrgyz Poverty Monitoring Survey	TSLS	-	Two-Stage Least Squares
LP	-	Linear Programming	UCA	-	University of Central Asia
LSS	-	Living Standard Survey			
LSMS	-	Living Standards Measurement Study	VRS	-	Variable Returns to Scale
ML	-	Maximum Likelihood	WB	-	The World Bank
MLE	-	Maximum Likelihood Estimates	WHO	-	World Health Organization

PREFACE

Acknowledgments

I would like to acknowledge my scientific supervisor, Prof. Dr. Peter Winker for his valuable comments and useful recommendations in shaping the structure of this research project, and for additional support. I am very grateful to the Central Asia Faculty Development Program of DAAD / UCA for selecting me, and sponsoring my study and stay in Germany, as well as the field research. I also want to thank S. Umetaliev, Head of the Organization and Territorial Management Department in the Government of the Kyrgyz Republic, whose support was very helpful in the field survey. Additionally I'm very grateful for the patience and skill of Mr. Arnd Federspiel, whose support in the correction of the English incredibly improves the perception of this thesis. I would also like to thank my family for inspiration, support and motivation, especially my wife Janna, who did the majority of the data processing, besides other support.

Kanat Tilekeyev

Germany, Giessen, July 2012

1. Introduction

Poverty and mainly rural poverty, is an important problem in Kyrgyzstan. Poverty in the country definitely existed in the pre-transition period, but the fall of the Soviet Union led to a sharp increase of the poverty level in the 1990s. The following decline of poverty, in the last decade of transition, is based on the recovery of the rural sector, growth of the trade in Central Asia and changing patterns of labor migration. Nevertheless, poverty reduction was not uniform and not stable due to an overall economic decline in the world and an increase of the main food prices. Rural remote areas in the country suffer from weak growth, a low capacity of the financial resources and the lack of human capital. The poverty level in those areas is volatile and steadily higher than the average trends in the country. This work is an attempt to study the determinants of poverty in the rural remote areas of Kyrgyzstan.

Sensitivity of the poverty rates to such variables as agricultural labor and overall economic production indicators (GDP, agricultural output and average wage) was detected on a country and a regional level. The linkage of poverty with the agricultural production capacity in the rural areas was selected as a working hypothesis. Further study of the production theory developments brings us to a separate area of the production function methods known as *frontier* approach. The main idea of the frontier theory is an explanation of the difference among producers, whose investments consist of equal inputs, but who obtain a different production output, due to its different technical inefficiency. Such an approach fits our purpose - technical inefficiency may explain the differences between the rural households' poverty rates. It is necessary to note that the typical rural household in Kyrgyzstan, as a result of the land reform implemented at the end of 20th century, became an independent agricultural producer with a separate small land plot and livestock. Currently, 95% of the agricultural output in Kyrgyzstan is produced by small-scale farmers and the rural households.

To validate the proposed poverty-efficiency linkage, three stages of research were arranged. At the first stage, the theoretical approaches of the poverty measurement and frontier methodology framework were defined. The second stage of research was the organization of a survey aimed at collecting appropriate primary data on a household level. This data gave us the chance to measure the poverty level and production efficiency estimations of the sampled households in the selected target region of Kyrgyzstan. A processing of the collected data, poverty measurement and technical efficiency estimations, and, finally, the modeling of the variables correlation were conducted during the final stage of research.

The thesis is presented in the five following chapters. Chapter 2 shows the specific features of the country, explaining natural resources, geography and climate conditions, historical and political background, and poverty trends, and is finalized by the formulation of re-search questions. Chapter 3 gives an overview of the poverty measurement methodology, including specific procedures of poverty measurement in developing countries, and poverty measurement practices in Kyrgyzstan. Apart from that, there is a separate description of the production frontier methodology development and different methods and conditions of the efficiency estimations. Chapter 4 is dedicated to a description of the process of data collection in the field survey in the selected region. Chapter 5 presents descriptive statistics of the sample - demographic, production and income-related parameters. The last chapter, Chapter 6, is dedicated to the final stage of our research - poverty and production efficiency measurement and modeling of the poverty-efficiency linkage.

As stated above, Chapter 2 describes important peculiarities of the Kyrgyz state in three spheres - the country's natural conditions, historical, political and economic background, and general poverty trends. Its highly mountainous nature defines significant differences of climate, remoteness of mineral resources, and scarcity of land available for agriculture. A historical overview describes the roots of the Central Asian backwardness, and the definitive role of the Russian Empire's colonization policy in the region, as well as the Kyrgyz develop-ment within the Soviet Union's structure. The political and economic transformation during the more recent history of the country highlights the contradictive character of the devel-opment. The poverty trends described in this chapter reflect the Kyrgyz society's reaction to the cardinal post-communist transformation and the following adaptation to new market conditions. The chapter ends with the formulation of the research questions.

Chapter 3 is divided into two separate parts - poverty measurement and production fron-tier. The description of poverty measurement covers a general overview of the basic poverty measurement pillars. It's also includes a description of the international practice of poverty measurement in the Third World, actively promoted by the World Bank (WB), and the poverty measurement methodology and its practice in Kyrgyzstan. Of the different pro-duction frontier approaches, Data Envelopment Analysis (DEA) and Stochastic Frontier Analysis (SFA) are most frequently used. The first method is a non-parametric deterministic approach to a production function, while SFA is a parametric stochastic estimation method. A description of both models and important features is given, as well as a discussion of the appropriateness of the presented methods' application.

Chapter 4 is dedicated to the description of field survey implementation in the selected region. The main forces of the selected region are presented. The data collection methodology includes a detailed description of the research questionnaire, sampling methodology and survey organization. The questionnaire is based on the examples of the Living Standard Survey (LSS), implemented in the country with support of the WB in the 90s, and on the questionnaire of the Kyrgyz Integrated Households Survey (KIHS), which is currently used for defining poverty rates. The sampling procedure is based on the geographical randomization selection of the households with the use of maps of the selected villages of Talas Oblast. Staff training, quality control procedures, project schedule and final results of the survey are given.

A descriptive overview of the sample is presented in Chapter 5. It covers a demographic overview of the sample including gender, age and social status pattern. Agricultural production of the rural households is explained separately, using two basic types of production - crop production and livestock breeding. The main parameters of crops and livestock types are presented. Apart from agricultural production activities, other income sources are also presented, including income from constant and seasonal employment, social transfers, small business activity and remittances.

The final chapter, Chapter 6, presents the results of the study. The poverty measurement part consists of the aggregate consumption construction method, adjustment of the poverty line, and finally the poverty results. The technical efficiency assessment is divided into two separate blocks - DEA and SFA application to the collected production data set. Traditional poverty measures are oriented towards the assessment of the whole population's poverty level, but are not applicable to separate households. The Poverty Gap Index was adjusted for the purpose of our research, in order to be applied to all sampled households. A proposed poverty index modeling shows a steady linkage with the some of the technical efficiency estimations of the sampled households determined through DEA and SFA application. The results of the analysis will be used for the final conclusions and policy recommendations.

2. Kyrgyzstan - Country Profile

This chapter describes the country background and offers a descriptive information review, supports clarification of the main features of the Kyrgyz economy and specifies the general trend of the poverty dynamics in the country. The chapter ends with the research problem formulation.

2.1 Geography, Climate, Natural Resources

Kyrgyzstan is a small country in the middle of the Eurasian continent located along the Tian-Shan and the Pamir-Alay mountain ridge. The country is known as the Kyrgyz Republic and has borders on Kazakhstan to the north, on China to the east and south, and on Uzbekistan and Tajikistan to the west. Its area totals close to two hundred thousands square kilometers (Abazov, 2008, p4).

Mountains occupy two thirds of the territory of the country. More than 40% of the territory is located on an altitude higher then 3,000 meters above sea level. Mountains separate the country from its neighbors to the south and southwest. The majority of borders lies on mountain ridges. Mountains divide the country into several parts. This factor complicates the communication and transportation between the different parts of the country. Only 5.1% of the land is covered by mountain forests, and 4.3% by water (Abazov, 2004, p2).

The climate in Kyrgyzstan is high continental and differs depending on altitude and region. The subtropical climate in the southern areas contrasts with the temperate conditions in the northern zone. The lower mountain areas have a dry continental climate, due to desert-warmed winds from Kazakhstan and Uzbekistan, while higher mountain zones have a polar climate. In the valleys, the average daily temperature in summer is around +30°C. In winter, the daily average temperatures are close to -15°C. Conditions are much colder at high elevations, where the average daily temperature in the summer is around +5°C and -28°C in the winter period. Precipitation differs between 100 and 500 millimeters in the valleys, depending on the region, and 180 to 1,000 millimeters in the mountains. Due to the dry climate, the agricultural activity has mainly been narrowed down to the irrigated lands, which only consist of approximately 7% of the country's territory. Except for irrigated agriculture in the valleys, almost half of the territory (around 45%) is taken up by mountain ranges (Abazov, 2008, p3-4).

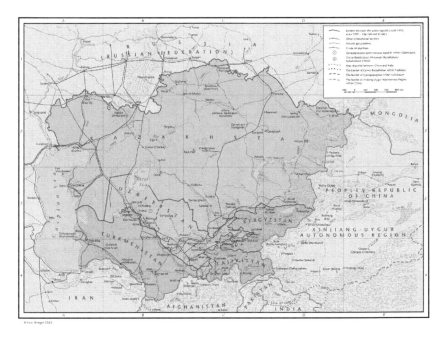

Figure 2.1: Kyrgyzstan on a Map of Central Asia, 2000
Source: Bregel, 2003

The country is rich in water resources. The country's hydroelectric potential is high, however it is only used partially. A lot of different mineral resources can be found in Kyrgyzstan , but the majority of these deposits are located in the remote high mountain areas, and big capital investments are required to explore them. Some of them, like gold, or coal and, in former times, uranium, have been actively exploited since the Soviet period despite the damages to the environment which continue to be caused until today (UNDP, 2005, p38, 84, 117).

2.2 History, People, Economy and Political System

The area of modern Kyrgyzstan is one of the most ancient centers of human civilization in Eurasia. Historians, mainly Russian and Soviet researchers, support the idea that the Kyrgyz nation originated from ancient nomadic groups migrating from the territory of Southern Siberia and Mongolia to Central Asia. The Kyrgyz nation settled on present-day Kyrgyz land during the 13th to 14th century. The devastating Mongol invasion in the 13th century increased the survival necessity of a nomadic tribal structure of the Kyrgyz ethos for a long period. The expansion of the Khanate of Kokand covered a significant part of Central Asia including the main part of southern and northern Kyrgyzstan during the 15th - 17th century (Adle et al., 2003, p110-114).

Beginning in the mid-19th century, the aggressive expansion of the Russian Empire in Central Asia, in contrast to British influence, led to a military and diplomatic colonization process. It ended with the inclusion of the territory of Kyrgyzstan, together with all neighbor territories, in the Russian empire at the end of the 19th century. As a part of the Russian Empire, Kyrgyzstan gained from important progressive innovations: internal military conflicts decreased; the custom of slavery was prohibited; public schools were opened; culture and natural resources were studied and the construction of roads initiated. Nevertheless, since the beginning of the 20th century, the further expansion of the Russian power was accompanied by a number of conflict situations within society. The majority of productive lands in the northern and central valleys was transferred to Russian and Ukrainian farmers who had been relocated from the western and central part of the Russian Empire. An increase of the tax burden, high corruption and the introduction of labor mobilization of the Kyrgyz, due to the participation in World War I, led to a military uprising in the whole of Central Asia, basically in North and Central Kyrgyzstan and South Kazakhstan. The uprising was bloodily suppressed, and the repression was stopped only at the beginning of the Russian Revolution in 1917 (Adle et al., 2005, p37, 40, 82, 94, 147,265-267).

After the Russian Socialist Revolution of 1917, Kyrgyzstan as a part of the former Russian territory joined the Soviet Union and developed further under its control (Everett-Heath, 2003, p108). It needs to be pointed out that a significant number of Russian migrants had already settled in Central Asia. Nominal autonomy was established in 1924 and the status of the Kirgiz Soviet Socialist Republic was nominally granted in 1936 within the state reform arranged by Stalin (Abazov, 2004, p21). The Kyrgyz gradually converted from a nomadic to a sedentary lifestyle, while the country was rapidly industrializing and up to a certain level urbanizing. Nomadic farming gradually declined as collectivization took place. Since the end of the 19th century constant migration of the ethnic Russian and Russian-speaking nationalities was observed in the area in several waves of the migration process. However, significant investment in the infrastructure, education and health care system accompanied the migration processes in the Soviet period. Soviet rule also led to the development of a specialized military-oriented industry in the whole Central Asian region. It is necessary to specify separately that all the Central Asian Soviet republics, except Kazakhstan, developed within the Soviet planning system as one integrated region, both from infrastructural, energy and security points of view (Everett-Heath, 2003, p109,111,147).

The end of the Soviet era was followed by the inability of the late Soviet leaders to control the situation. Together with the other post-communist countries, Kyrgyzstan also experienced

Figure 2.2: Administrative Map of Kyrgyzstan, 2001

Source: Map No. 3770 Rev. 6 United Nations, Department of Peacekeeping Operations, Cartographic Section, Jan. 2004

most of the catastrophic distortions in structures, institutions, incentives and prices, due to the destruction of its economy and political system (Aslund, 2007, p28). The economic decline of former systems was caused by serious economic problems of the planned economy, which immediately negatively affected the well-being of the population (WB, 2001, p2). A country with active foreign support in form of credits, grants and technical assistance implements a number of market reforms with the aim of liberalizing the economy, normalizing the balance of payments, privatization and land reform (Everett-Heath, 2003, p120; WB, 1999, p1). The main parts of the reforms were implemented from 1993 to 1999 (Abazov, 1999b, p218-221). The private sector grew slowly in the 90s, but started to be active and important in the more recent past. As a result, now around 90% of all industrial and agricultural products are produced by the private sector (UNDP, 2006a, p12).

The GDP (Gross Domestic Product) regrew slowly in the 1990s after a dramatic decrease due to the collapse of the Soviet Union (Kort, 2004, p162). At the same time, the economy's structure changed too. The GDP structure changed to more income from the service sector

(44% compared to 33% in 1991). Simultaneously, agriculture remained on the same level after the significant growth in the 1990s, but the role of the industry seriously declined (Aslund, 2007, p167).

The main part of the population (close to two thirds) in the country lives in two geographically separated valleys - Chui (in the north) and Fergana (in the south). The country population consists of 5.5 million inhabitants (end of 2010). Almost two thirds of the population are located in the rural regions. 80 different nationalities and ethnic groups live in the country, but the majority group of Kyrgyz people consists of 3 nations only: Kyrgyz (71.7%), Uzbek (14.4%) and Russian (7.2%). The Uzbek population is concentrated in the southern regions, in the Fergana valley neighboring on Uzbekistan, and the main part of the remaining Russians in the Chui valley and the capital Bishkek in the north of the country (NSC, 2011a, p18-103).

Migration processes in the country have been characterized by a persistent outflow migration since the beginning of the 90s. But character and intensity of the external migration changed strongly over last two decades. In the first half of the 90s, the republic was faced with an extremely high outflow of the Russian population, mainly to Russia, including educated and skilled professionals and specialists (Abazov, 1999a, p246).

Over the next five years the migration processes significantly decreased and have risen again since 2001. The main destination of migration is Russia and, on a lower level, Kazakhstan. There has been an annual net outflow migration of around 0.7% of the population within the last five years. The new trend in emigration increases the share of ethnic Kyrgyz among those leaving the country to up to 50%, others are mainly ethnic Russians and Uzbek (NSC, 2011a, p306-346).

Kyrgyzstan as a state has a unitary system, and consists of seven regions (Oblasts): Issyk-Kol, Naryn, Osh, Jalal-Abad, Batken, Talas and Chui and the two big cities Bishkek (capital) in the north (around 0.8 million inhabitants) and Osh (around 0.4 million inhabitants) in the south. Before 2000, Osh was included in the Osh Oblast. Oblasts are divided into a total of forty districts (Rayons) (Abazov, 2004, p53-54).

The latest period of the development was characterized by a number of negative trends, and threats increased in the country. Low state control and a high corruption level has led to a significant growth of the informal sector, which was assessed at around 53% of the official GDP in 2006 (UNDP, 2006b, p3). The high level of labor migration to Russia led to an increase of remittances and to a loss of labor force, including educated and skilled workers (Ibragimova, 2008, p33-34).

High level of corruption (OECD, 2005, p15) and low democratic institutional environment negatively affected the civil development of the country and resulted in 2005 in the first revolution (ICG, 2005, p2). However democratic reforms initiated in the period after 2005 was not realized and tension in the society increased (McMann, 2006, p181-183). The political process in 2005-2008 consist of the political power concentration in the hands of the second President Bakiev. Main results of that trend was full control of the parliament and court system, prohibition of the opposition media, starting of the corrupted sale of the energy sector and bringing his close relatives to the power. At the same time consumer prices in 2008 raised by 25%, while food prices raised to 32%. As a consequence, around 1 million poorest people in Republic became vulnerable to higher food prices (Martino et al., 2009, p24,26). Hidden contradiction of the opposition rose and supported by the wide population dissatisfaction finally finished by the militant resign of the President Bakiev in April 2010.

This situation of the state's loss of control led to an aggravation of the ethnic conflict between the Kyrgyz and the Uzbek population in the southern region in June 2010. Around 500 people died in this conflict. It signalized how deep and serious the contradiction between the dominating growth of the Kyrgyz ethos influenced society and the inability of the state to arrange a fair and equal access to power for the Uzbek diaspora in the country (KIC, 2011, p44).

The newest political arrangement in the country is a democratic-oriented parliamentarian system. It was introduced instead of the strong presidential system. Third President Roza Otunbaeva, elected soon after the ethnic conflict in June 2010, came temporarily into power for the transition period of 2010-2011. Otunbaeva arranged a full support for the democratic election of the new parliament in 2010, which included all the political forces in society. The end of 2011 saw the election of new President Almazbek Atambayev. Despite some imperfections detected by international observers, the election was assessed as transparent and legislative. The political system has stabilized and signalized the completion of the transitional period after the 2010 events (OSCE, 2012, p3-4,20-23).

2.3   Poverty Trends in Kyrgyzstan

The poverty issue was basically ignored or hidden in the economic analysis during the socialist period of the country due to Soviet ideology, but has become an important issue since the coming of independence (WB, 1995, p1). Therefore we need to describe here shortly two separate stages of the issue analysis - the pre-transition socialist situation and latest trends starting at the beginning of the 90s.

Historical records concerning the poverty issue for the socialist period of Kyrgyzstan are quite poor on the one hand, and contradictory on the other. Researchers clarify that in the pre-transition period an extensive system of social protection was established in the country, which was also complemented by a developed social service infrastructure including free education and a health care system (WB, 1995, p1,12).

However, poverty was indicated in the society. It was formally called inability to meet a socially acceptable standard of living. The concept was developed in the mid-60s by Soviet researchers Sarkisyan and Kuznetsova. The term "poverty" was avoided in the Soviet research literature, and was finally replaced by the word "underprovisioned". The concept provides a certain 'poverty line' formed by the fixed rate of the minimum income level per capita. It was based on the creation of the typical budget for the average Soviet citizen. The budget includes a basic set of food and services typical for the Soviet consumption pattern. This set was unified for the whole Soviet Union. The budget finally defined as minimum monthly income 51.9 Soviet rubles for each person per month (Matthews, 1986, p17-22). Literature further used the rounded sum of 50 Soviet rubles per person. Correction on inflation or population strata patterns were not provided. In the first poverty assessment in Kyrgyzstan, a poverty line was fixed at the level of 75 rubles per month per person. Those persons who received less income were considered as living poorly and lacking in supplies (WB, 1995, p12).

Available data from the Soviet Household Budget Survey (HBS) on income distribution shows that in 1989, 35% of the people in Kyrgyzstan received a monthly income of less then 75 roubles per month. Food expenses was dominant in the consumption pattern of poor people. People with low income concentrated in the rural areas and tended to have bigger families with a high rate of dependants. Finally, it was concluded that at the end of the Soviet period in Kyrgyzstan at least one third of the population was poor (WB, 1995, p12).

However, data for household budget surveys were derived from an unrepresentative survey system. It was based on enterprize work-force rolls. Therefore, certain groups of people, e.g. retired persons, peasants from small farms and officials and military forces too, were not covered by this survey (WB, 1995, p12). Another important factor which needs to be mentioned here that in the income was not included the unofficial activity of people, including their own food production, supports an additional unregistered income source. Thus, poverty clearly existed in the country in the pre-transition stage, but a conclusion on a more or less accurate level could not easily be drawn due to the absence of an objective measurement system.

Poverty measurement based on an objective methodology was first arranged in 1993 under World Bank (WB) guidance. The first national survey was based on the Living Standards Measurement Study (LSMS) methodology developed and used by the WB in developing countries since the beginning of the 80s (Chander et al., 1980, p1-3). This survey plays a fundamental role in defining the principles, methods and practices of all surveys of such type. A more detailed description of the main methods and tools used for poverty measurement and important for the current work will be described further in the part dedicated to the theoretical methodology description.

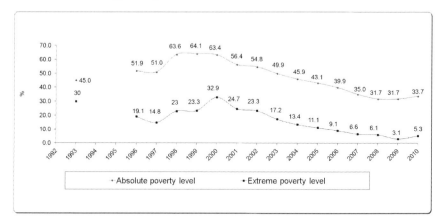

Figure 2.3: Poverty Trends (Headcount Index) in 1993-2010
Source: World Bank (1996-2008), National Statistical Committee (2001-2011)

From 1996 onwards, the conduction of surveys became a regular procedure, but the support-ing donor agency, the format of the main characteristics, methodology and samples changed several times until 2003. Surveys were implemented by a team formed by the National Statis-tical Committee (NSC) under the supervision of the World Bank and other donor agencies. Since 2000, surveys have been fully implemented by the NSC, however technical support from the WB and the UK Department for International Development (DFID) has been kept (WB, 2007a, p5,18).

National poverty indicators in aggregated form may be seen as a more or less comparable time series line over the observed period, because most of the surveys are based on LSMS methodology, which provides the constancy of the basic methods of survey. Modifications mainly concentrate on the adoption of the questionnaires, decrease the mistakes in surveys, change the methods of the covering respondents' base, and widen the base of collected data to cover some of the issues missed at the initial stage. However, in many aspects (starting from the design of the expenditure module and finishing with the size of the sample) different household survey instruments do not allow us to use the poverty data for a deeper analysis of the poverty trends over time or for making certain generalizations for the specific analysis of the poverty of certain subgroups (e.g., ethnic, educational, or gender groups) within certain regions or across the surveys in certain periods of time (WB, 2005, p18-20,23-27).

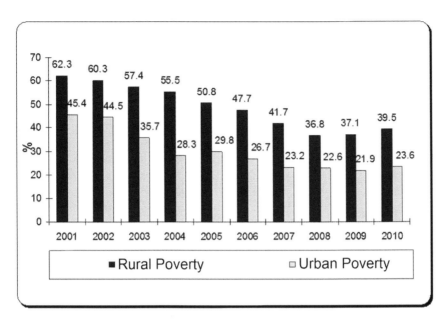

Figure 2.4: Rural and Urban Poverty (Head Count Index) in 2001-2010
Source: National Statistical Committee (2001-2011)

Poverty dynamics in Kyrgyzstan in the 90s were characterized by a high increase of the poverty level. The highest poverty level was recorded in the period 1998-2000 - during that time, almost two thirds of the population lived below the poverty line. Other poverty indicators also reached maximum level. The poverty analysis shows a strong correlation with some geographical, demographic and educational variables (Pomfret and Anderson, 1999, p4-8), which are still valid today:

- Poverty is mainly a rural phenomenon, and constantly higher in the southern regions; the lowest poverty rates are recorded in Chui Oblast and the capital Bishkek.

- Households with a large number of children are poorer, predominantly Kyrgyz and Uzbek households in the south, however the ethnic effect diminished further in the 90s.

- In general, the education of the household head is not significant below the tertiary level, but having a household head with higher education decreases the probability of being poor.

From 2001 onwards, poverty in Kyrgyzstan started to decrease rapidly. It reached a pre-

13

transition level of approximately one third of the poor in society in 2007. However, since that year the poverty indicator stagnated at a level of around 32% over the next two years and in 2010 started to increase slightly. The main reasons for the decrease in poverty are an overall growth of the economy (WB, 2007a, p26), the beginning of the emigration of labor to Russia and Kazakhstan and the following remittances increase (Ibragimova, 2008, p68-72), an increase of regional trade activity (WB, 2007a, p28) and a modest growth of agricultural productivity (WB, 2007b, p25-26). Despite a serious decrease of poverty trends, the issues of stagnating poverty and the absence of further sources of poverty alleviation still arise in the area. The latest negative economic and political threats specified earlier (e.g., high inflation, political instability, inter-ethnic clashes) already start to deteriorate the latest improvements and may certainly lead to a number of problems for the future development of the country.

2.4   Research Question Formulation

The analysis of poverty dynamics in the country was done previously as part of the preparation of the project (Tilekeyev, 2010, p48-51). Modeling of poverty with macroeconomic indicators shows that poverty is sensitive to two important factors:

- positively affecting overall economic indicators - GDP, agricultural output in rural regions, average wage;

- negatively affecting economic indicators - agricultural labor.

The time series models were consistent on the national and regional levels, but due to short historical lines they (1996-2008) cannot demonstrate a solid evidence of the main poverty determinants. However, the results of the empirical correlation of poverty trends in Kyrgyzstan bring us to the conclusion of a possible linkage between poverty and economic performance of society and labor mobility in the rural sector. In more specified terms this idea means that the poverty level may be influenced by the production efficiency in the rural areas, as it is known that poverty is predominantly a rural phenomenon. Another factor supporting this idea is high sensitiveness to labor dynamics in agriculture, which is also related to the basic elements of the production theory and specifically the production function theory. It studies the effects of labor and capital optimization on the final output in different sectors, countries and periods of time. Naturally, conclusions regarding the agricultural production efficiency as an important poverty determinant become a main idea in the research.

In order to narrow down the problem we prefer to specifically concentrate on studying poverty in the remote rural areas as especially vulnerable parts of the country. The main research

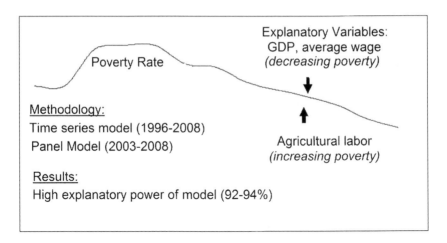

Figure 2.5: Poverty Determinants Modeling
Source: Author's own presentation

questions we want to address in this research are:

- What are the poverty determinants in the remote rural areas of Kyrgyzstan?

- Does a linkage between the poverty and efficiency of the rural households exist in the remote rural areas of Kyrgyzstan?

- Where are the limits of rural poverty improvements in the remote areas of country?

In order to answer those research questions we examine the potential linkage of the poverty indicators in one selected target remote rural region of Kyrgyzstan through studying the effects of production efficiency on poverty. The hypothesis can finally be formulated as follows: "Poverty in the remote rural communities of Kyrgyzstan may be determined by the limits of the agricultural production possibilities".

3. Theoretical Background

This chapter consists of two blocks and describes the review of theoretical thought dedicated to poverty and production efficiency problems. The first part is a review of the poverty measurement thought, which helps to formulate the definitive part of the problem - the methodology of poverty measurement, international practice and standards, poverty indicators and measurement issues.

Separately analyzed is an approach to practical poverty measurement based on the specific circumstances of Kyrgyzstan. The second part is devoted to production function theory developments relating to measuring productivity and efficiency on a microeconomic scale. Two main frontier efficiency analysis methods are reviewed - the Data Envelopment Analysis (DEA) and the Stochastic Frontier Analysis (SFA), with their specific assumptions and specifications.

3.1 Poverty Measurement Methodology

3.1.1 Poverty Measurement Tools

The definitive part of poverty measurement approaches and methods is important as it is directly influenced the results of the empirical analysis of poverty. The description of the main components and parameters of poverty analysis helps us to concentrate our attention on the practical application of the poverty analysis and its linkages.

The main problem distinguished in the poverty analysis is the two-step procedure of *poverty identification* and *poverty aggregation* (Sen, 1976, p219-222). Thinking of the first problem, its basic dilemma is that of poverty subject parameters selection and poverty line selection and construction. The second problem emphasizes the importance of an appropriate aggregation tool (measures) for the poor households (or individuals) and its characteristics.

The discussion of poverty identification is mainly an issue of the definition of poverty and of the main parameters of the poverty application. Poverty in general is defined as the inability of an individual or a household to command sufficient resources to satisfy basic needs (Fields, 1994, p88). The needs mentioned in the poverty definition are understood as a set of basic necessities, including food, clothing, housing and basic services. Poverty identification requires the definition of the main poverty subject characteristics, which are the following (Haughton, 2005, p14-63):

- Choice of income unit to measure (e.g., household, family or dwelling), including necessary assumptions;

- Choice of the measured resources (e.g., income or consumption);

- Adjustment of selected resources for family size and composition: adjustment for economy of scale and for the adult equivalent in relation to children's consumption;

- Choice of poverty line. The poverty line establishes a certain level in such a way that households at or below the line are recognized as 'poor'. There are the following two types of poverty lines:

  [i] Absolute poverty line. This establishes the cost of the particular 'basket' of 'minimal' needs (set of goods and services) fixed as a certain sum of money. Every unit with an income under this line is counted as poor.

  [ii] Relative poverty line. This type of poverty line establishes the poverty line as a percentage of mean (or median) income in a selected society.

The whole concept of poverty definition concentrates on the question of measuring the household income and correlating it with the particular level of income which plays the role of the socially acceptable consumption norm. This correlation brings us to the final conclusion of the poverty rate, depth and characteristics.

Despite some criticism of the monetary measurement of poverty it is still the main instrument of poverty definition and measurement. The new developing area in poverty analysis is the multidimensional poverty approach, which is still in the process of establishing itself. Thus, when analyzing the well-being of Kyrgyzstan's population, it is skipped in this current work. Certain monetary parameters, income (or consumption) of the particular household (individual), which could be measured and analyzed with different tools, define certain characteristics. Such analysis and data are necessary in order to urgently define the strata of the population suffering from a lack of needs, and to correct the government's economic and social policy.

Aggregation discussion basically developed by the two important works of Watts (1968) and Sen (1976). Watts firstly formulate the definition for the measurement of poverty, and Sen constitutes axiomatic approach to poverty measurement. But Watts work do not receive an attention among poverty specialists, until rediscovering of his work in 80s by Foster (1984) and in 90s by Zheng (1993), while Sen's article has fundamental impact on the poverty analysis development.

Sen proposed three basic axioms a poverty measure should satisfy (Sen, 1976, p219-222) - *focus, monotonicity* and *transfer*. Defining that for the income space $\Psi = \bigcup_{n=1}^{\infty} \Psi^n$, any poverty line z, $z \in \Psi$, and any income distribution y, $y \in \Psi$, it is possible to construct a poverty measure $\mathbf{P}(y; z)$ (Ziliak, 2005, p30-31). Then axioms are defined as follows:

- *Focus Axiom*: changes of the income of the non-poor should not affect poverty: $\mathbf{P}(y; z) = \mathbf{P}(x; z)$, whenever $y \in \Psi$ is derived by $x \in \Psi$ by an increment to a non-poor person,

- *Monotonicity Axiom*: changes of the poor object income (consumption) should affect the measure even if the object continues to be poor. Two forms of the axiom exists - 'weak' and 'strong' monotonicity:

    (i) *Weak Monotonicity*: $\mathbf{P}(y; z) > \mathbf{P}(x; z)$ when $y \in \Psi$ is derived from $x \in \Psi$ by a simple decrease in a poor person's income,

    (ii) *Strong Monotonicity*: $\mathbf{P}(y; z) < \mathbf{P}(x; z)$ when $y \in \Psi$ is derived from $x \in \Psi$ by a simple increase in a poor person's income, even if y thereby crosses the poverty line,

- *Transfer Axiom*: transfer of income from a 'richer' poor person to a 'poorer' poor person should also be reflected by a poverty index. In this case if y is obtained from x by a regressive transfer then $\mathbf{P}(y; z) > \mathbf{P}(x; z)$. This definition is also known as the 'weak' transfer axiom, however at least seven other variants of this axiom were additionally developed later (Zheng, 1997, p82-83).

Traditional poverty indexes (headcount ratio, income gap ratio and poverty gap ratio) used in the literature fulfill these axioms simultaneously, and this fact motivated Sen to develop his own poverty index, and further intensive discussions on an 'ideal' poverty index led to the creation of a number of poverty indexes in the 80s and 90s. The two-step procedure, suggested by Sen, became a classical procedure in poverty analysis, while no unified agreement was reached about certain distribution-sensitive poverty index.

The decision of the axiom fulfillment were Foster-Greer-Thorbecke (FGT) family measures (Foster et al., 1984, p762-763), which became widely used by practitioners, primarily by the World Bank poverty specialists. At the same time, the Sen index was criticized for the lack of intuitive appeal demonstrated by some of the simpler measures of poverty, but also because of its inability "...to decompose poverty into contributors from different subgroups" (Deaton, 1997, p147), while FGT fulfilled this task successfully. Currently all poverty measures which are introduced belong to the FGT measures (WB, 2007a, p147).

Thus, the distributional measures of poverty include the three following indexes - poverty headcount ratio, poverty gap ratio and poverty severity index (Haughton, 2005, p69-75). Selected measures make up the mainstream approach in measuring poverty. The poverty headcount ratio or **Headcount Index** counts the persons with an income located under and on the border of the poverty line and calculates the proportion to define the level of aggregated poverty. This index, typically denoted by $\mathbf{P}_0$, is the most widely used in practice and can be defined as:

$$\mathbf{P}_0 = \frac{\mathbf{P}(y; z)}{\mathbf{n}(y)} \qquad (3.1.)$$

Where, $\mathbf{P}(y; z)$ is the number of poor $(\mathbf{y_i} < \mathbf{z})$ and $\mathbf{n}$ is the population size. The Headcount Index is characterized by a number of weaknesses: there is no information about the intensity of poverty, its depth, and the necessity to recalculate it with regard to individuals but not to households. However, it is quite simple to calculate and understand, which helps to check the general trend across countries and whole regions (Haughton, 2005, p69-75).

The next index is the **Poverty Gap Index**, which measures the percentage of the average shortfall of the poor persons to the poverty line. It is formally defined as:

$$\mathbf{P}_1 = \frac{\mathbf{P}(y; z)}{\mathbf{n}(y)} * \left(1 - \frac{\mu(y; z)}{\mathbf{z}}\right) \qquad (3.2.)$$

Where $\mu(y; z)$ is the mean income of the poor. The equation is a product of the combination of the Headcount Index and another index (part of equation 3.2.), known as *Income Gap Index*, which measures the percentage of the average shortfall of the poor to the poverty line. This index helps to see how far poor people in general are located from the poverty line, however it is still insensitive to the transfer axiom, which requires transfers across the poor population to be taken into account.

The last index is the **Poverty Severity Index**. This index belongs to the family of Foster, Greer and Thorbecke measures. It uses the weighting function and demonstrates the depth of poverty by giving a higher value for poorer people. To construct it, it is necessary to use the squared poverty gap index. Formally it is defined as,

$$\mathbf{P}_2 = \frac{1}{n} \sum_{i=1}^{n} \left(\frac{z - y_i}{z} \right)^2. \tag{3.3.}$$

This measure is not used very widely, but it is useful as it can be applied for population sub-groups' contribution to the disaggregation of the overall poverty in the country. A combination of the different poverty complementary indexes supports the overall poverty analysis.

Completing the main tools and characteristics of the poverty measures is a necessary step to clarify the perception of the problem from a purely technical point of view. The next step, described in the following section, is to define the correct and clear framework of the main source of the poverty analysis - household survey, as the primary data source of poverty measurement.

3.1.2   Living Standards Measurement Study

The Living Standards Measurement Study (LSMS) is a main poverty study instrument of the World Bank. It has been developed since the beginning of the 80s (Chander et al., 1980, p1-3). The main reason for creating an integrated unified structure for implementing household poverty surveys was the absence of a common platform for poverty measurement in developing countries. Significant experience concerning poverty measurement was collected starting from the 50s in Asia, Africa and Latin America. This huge amount of information, however, was characterized by a number of serious differences in approaches, concepts and an often different quality of data regarding the well-being in poor countries. Such a methodological inconsistency strongly deteriorates the formulation of an effective anti-poverty policy for a number of the less developed counties all over the world (Deaton, 1997, p32-34).

Another important problem concerned the weak and quite costly field data information processing logistics in the developing countries. It usually took several years after carrying out the surveys before the results were accessible, and the quality of the data was also a critical issue. The introduction of information technologies in the field conditions increased the speed of methodology adoption, improved the control of data quality within the process of the field survey implementation, data processing speed and data quality itself (Deaton, 1997, p36).

First LSMS surveys was implemented in Latin America (Peru) and Africa (Cote d'Ivoire) in the mid-80s. Currently, LSMS surveys are implemented in about forty developing countries. The surveys create new standards in the household surveys and seriously affect the current field practice. Until recently it was the only regular source of information about the poverty and well-being of the people in Kyrgyzstan. Alternative sources of such kind of information are still in the process of being formed and shaped and cannot be used as methodological and factual basis for the current study.

The Living Standards Survey's (LSS) prototype structure (Grosh and Muñoz, 1996, p5) suggests the six main elements critically important for the survey - questionnaire, sampling, field operations, data management, data analysis and planning program. The practical matter of interest for us would only be the first three elements, because the remaining three are basically oriented towards the creation of organizational frameworks for the local statistical offices of the countries where LSS was implemented. It has to be pointed out that LSMS is clearly an attempt of the World Bank to introduce a common practice of a clear and effective system of poverty measurement on an objective methodological platform.

Questionnaire: Three types of questionnaires were developed - a household questionnaire, a community questionnaire, and a price questionnaire. The main source of information is the *household questionnaire*. It consists of the different sections describing all spheres of household activity. It collects individual-level and household-level information. The typical list of sections includes the following items: household structure, dwelling characteristics, education data, health information, economic activities and employment specifics, migration, agricultural activities, nonfarm self-employment activities, food and non-food expenditures, durable goods expenditures and inventories, fertility, other income sources, savings, assets, and credit markets, and anthropometric measurement of household members (Ainsworth and van der Gaag, 1988, p6; Deaton, 1997, p35; Ravallion, 1992, p9).

An interview is usually split into two rounds, with a two-week interval. The first session is devoted to the collection of individual levels of information, while the second visit concentrates mainly on the expenditure module. Typically, the responsible household member should keep a diary of all expenditures made in the reported period. Such a scheme improves the quality of data and decreases the time of interview per visit. The typical time spent on filling in the questionnaire varies from 3 to 4 hours (Deaton, 1997, p26).

The *community questionnaire* is typically employed in rural areas and collects general information about a village - demography, climate, migration and infrastructure, transport, public services, water and energy supply. The linkage between household and community level allows us to analyze the access to different services or the impact of certain infrastructural projects on the well-being at the household level. The *price questionnaire* similarly collects the points made by people interviewed in every population. It is simply a list of products with the prices per unit. It helps to adjust living standards in the different parts of the country and in countries with high inflation, as a survey typically requires long periods of time (Ainsworth and van der Gaag, 1988, p67-69; Ravallion, 1992, p16).

Defining of a questionnaire is a multi-step procedure, which passes several discussions of revisions with the involved stakeholder from the local government, field tests of the questionnaire and additional revision based on the results of the field test. A typical event at the finalization of this process is a seminar describing the questionnaire and the further planning procedure. Iterative processing of the questionnaire development helps to improve and adopt it to local conditions. Normally, this process takes several months and requires serious human and financial investments (Grosh and Muñoz, 1996, p22-28).

Sampling: The Living Standards Survey is a relatively small-sized survey and typically varies from 2000 to 5000 households. The sample normally represents the whole population of the country, but in special cases excludes people living in areas of local conflict, migrating nomadic groups, etc. The sample should also represent subgroups of the population, e.g., rural-urban subgroups, ethnic minorities (Howes and Lanjouw, 1997, p3-8; Ravallion, 1992, p11-12). Design of the sample normally consists of a two-stage procedure:

- The first stage consists of the selection of Primary Sampling Units (PSU) across the country.

- At the second stage, a certain number of households is selected in each PSU.

Both of these stages are random selections. The first stage of sampling requires the creation of a sample frame from census data and at the second stage a full listing of census data for selected PSU is necessary for the selection of final random samples. The two-stage procedure reduces the cost and effort compared to the single-stage sampling procedure, but increases sampling errors. Unbiased estimation from the survey is overcome by using weighting values for observations in the survey.

It is important to additionally specify four concepts of the sampling theory applicable for LSMS: sampling error, non-sampling error, multi-stage sampling and analytical domains (Grosh and Muñoz, 1996, p54-60).

The *sampling error* is the error originating from observing part of the population and attempting to make inferences about the population as a whole. The first rationale this needs to be reconciled with is the law of diminishing returns, which means that the increase of sample size does not seriously affect the error. The second reason is that sample size precision in general is almost independent from the total population, and different sample size basically arises from the necessity to produce poverty indicators at the regional level.

Non-sampling errors are known as errors originating from natural causes - refusals, respondent fatigue, interviewers' errors, or the lack of an adequate sample frame. Non-sampling errors are unpredictable and could be decreased through good organization of the survey.

Bigger sample size typically leads to higher non-sampling errors (Grosh and Muñoz, 1996, p56).

Multi-stage sampling is the tool to decrease costs of the sample selection. In case of single-stage sampling, access to the national list of households is required and transport costs are also increased significantly. The typical two-stage procedure is supposed to select a certain previously decided amount of small area units with Probability Proportional to Size (PPS), then randomly select a fixed number of households in the area, with the same chance to be selected. Area units are the smallest geographic units in the national census, typically named Primary Sampling Unit (PSU). These PSUs, however, may be quite big and in this case the PSUs are divided into segments, and one of them is selected to represent the PSU (Grosh and Muñoz, 1996, p54-60).

Advantages of the multi-stage procedure are as follows (Grosh and Muñoz, 1996, p57-58):

- Each household in the PSU has roughly the same chance to be selected, i.e. the sample is approximately self-weighted.

- Transport logistics of the process are optimized in certain areas instead of distributed randomly across the country and the work load per interviewer is also better, as the number of households per PSU is similar

Disadvantages of this approach include a higher error level because it is more probable that neighboring households demonstrate similar characteristics than a simple random sample of the same size. As a result less population diversity is reflected by the sample. This influence is called *cluster effect* and originates from the *design effect* (Kish, 1965, p154-156,161). It arises with the decreasing number of PSUs and the increasing number of selected households in every PSU. An increase of the error level is normally observed within the range of 0.01 to 0.1, but in some extreme cases it may increase to 0.5 (Grosh and Muñoz, 1996, p59).

Analytical domains: For a number of reasons, which are mainly political, it is necessary to define certain groups of the population separately. Typical examples include the stratification of the urban and rural population and regional administrative subdivisions. However, certain subgroups may not be provided by geographical location, for example certain ethnic minorities, gender groups, heads of households employed in the public sector in urban areas, etc. In this case, the sample should be designed in such a way as to cover at least a minimal number of representatives of such subgroups. Such a design is called *analytical domain* (Grosh and Muñoz, 1996, p58). Basically, a large domain is automatically included in the sample, however normally the technique of oversampling a certain domain is applied in other cases, for a smaller domain, which is followed by the utilization of the modification of the expansion factor by introducing "sampling weight". The two-stage sampling procedure is applied independently within all imputed do- mains. At the same time it is impossible to include all the possible domains due to the risk of developing a prohibitively large total sample.

Field Operations: Here the prototype 'ideal' scheme of the field stage of the LSMS survey is introduced (Grosh and Muñoz, 1996, p83). In practice the organization of work typically changes depending on the country specifics. The field operations stage is conducted by a number of independent teams. Each team includes a supervisor, interviewers, data entry operator and driver. The team's work is usually arranged on an annual basis. Monthly teams cover two PSUs. Each interviewer visits eight households per PSU. The household questionnaire is filled in during two visits conducted in two-week intervals.

Data are processed continuously after each round. The reason for this procedure is to detect errors from the first visit and correct them. The supervisor controls the implementation of the work, including check-up interviews and processed data analyses. The central supervision of the process consists of a monitoring situation and of controlling the intermediate results of the work.

The advantages of such an organization of work is an increase of the collected data quality and an increase of the speed of data processing. The disadvantages include high requirements to the field workers and long periods of field work.

Thus, LSMS is the main source of the poverty measurement information in a number of developing countries, including Kyrgyzstan. The next section analyzes the process of introducing the survey in the country, the adaptation to local conditions, the further evolution of poverty surveys and the main peculiarities of LSMS and of poverty indicators methodology implementation in the country.

3.1.3   Poverty Methodology History in Kyrgyzstan

The history of the current poverty measurement methodology starts in 1993 with the first LSMS Survey, supported by the WB. The basic features of this survey created the particular pillars for measuring poverty in the country. Since this period were applied:

- an absolute approach in defining the poverty line;

- a basket approach with foodstuff plus non-food expenditures;

- there are two poverty lines - the poor and the extremely poor (severely poor), arising from the basket approach;

- poverty measurement is based on consumption expenditures;

- expenditures do not include the full cost of the purchase of durable goods;

- Headcount, Poverty Gap and Poverty Severity Indexes were introduced as main poverty indicators.

First, the construction of the *poverty line* was faced with the dilemma of choosing between the absolute and the relative approach. The absolute approach was selected because with this approach it was possible to define reasonable and acceptable norms of the minimum standards of living, based on the primary needs of a person (Ravallion, 1992, p25). At the same time, the relative approach is always discussible with regard to an acceptable minimum of needs. The rationale was described as follows: the main task was to define the people with the highest needs, then those who are relatively disadvantaged, but can still afford basic needs.

The main idea of defining the poverty line was to estimate the cost of a basket of goods, reflecting the basic needs (Ravallion, 1992, p26; Ravallion, 1998, p15). This method includes several stages. First, individual dietary intake data from a survey are collected and analyzed. The collected data form an approximate minimum food basket. This basket should reflect typical Kyrgyz food and at the same time should contain a certain calorific level. The norms of the food basket depend on age and gender and are based on the requirements of the World Health Organization (WHO) and the Food and Agriculture Organization (FAO), adjusted for the mean weights of the population (WB, 1995, p2,50).

Based on the collected data two baskets are calculated. Both baskets pro- vide the required level of nutrition and allow an adequate growth and activity. However, the content of these baskets differs. The "low" cost basket includes a more austere diet and reflects the current consumption pattern of the poorest population in Kyrgyzstan. The "high" cost basket reflects the mean figures typical for average consumption (Ackland, 1996, p38).

The collected data showed that the low income population spent around 80% of expenses on food and only 20% on the non-food stuff and services. This ratio was taken as the base for calculating the poverty line. The food baskets were increased according to this share and became the first poverty line. The "high" basket became an absolute and the "low" an extreme poverty line. For each household the poverty line was calculated as a sum of the "baskets" for all the household members (WB, 1995, p50).

In 1996 it was detected that the poverty line did not adequately reflect changes due to the huge structural shift of the relative prices. The new poverty line was recalculated in 1996 (WB, 2001, Annex 1). The next recalculations of the poverty line were in 2003 and 2008.

Since 1996 some changes were introduced in the calculation of the poverty line and the next readjustment was made in 2003. First of all, the energy intake value was fixed at a level of 2100 calories per capita per day, by recommendation of the WHO. The extreme poverty line was calculated on the basis of the energy value and includes food items only. To fix the extreme (food) poverty line, a reference group in the population was identified. This reference group was set as the third, fourth and fifth decile of the population. The group's food pattern was analyzed. 97 food items were selected from 579 of all food and beverage items listed in diaries. These core food items constituted 97% of all food consumption for the target population, reflecting a relatively low-income population (WB, 2005, p16).

The absolute poverty line was calculated as the sum of non-food items and services added to the first calculated extreme (food) poverty line. The upper-bound method was adopted to determine this line (Ravallion, 1992, p27-29; WB, 2005, p49):

- The reference group included the first three quarters of the population (quartiles 1, 2 and 3);

- the share of non-food expenditures for the reference group was calculated (39,8% in 1996, 37.1% in 2003);

- this share was added to the value of the food poverty line to calculate the absolute poverty line.

Table 3.1: Poverty Line Dynamics in Som and US PPP Dollars

		1993	1996	2003	2008
1	Absolute poverty line				
a	Som per person per capita	1048	3652	8732	18323
b	US $ PPP per person per capita	1545	868	927	1288
2	Extreme poverty line:				
a	Som per person per capita	659	2199	5490	11710
b	US $ PPP per person per capita	971	522	583	823

Sources: (WB, 2007a; NSC, 2010b; Heston et al., 2012)

The last recalculation of the poverty line was made in 2008 by the National Statistical Committee, but a more detailed explanation of the methodology was not found in the open sources. However, there is a high possibility that the last adopted methodology will continue to be used. All the poverty lines presented above are shown in Table 3.1, presenting them in both national currency and the equivalent US dollars in Purchasing Power Parity (PPP).

The expenditure approach was chosen as the primary indicator of welfare for measuring the poverty level. The following advantages of expenditures were taken into account in contrast to the income measures (WB, 1995, p49):

1 If data are based on the income approach only, this does not take into account the possibility of households' use of savings or borrowing and of inter-household transfers such as gifts or support in cash and aid in kind from friends and relatives.

2 The use of income data does not include underreporting for income received from the private and informal sectors.

3 The income of a household may fluctuate strongly within a short period of time, which is typical for rural areas due to strong seasonality effects.

Personal consumption expenditures include the following consumption aggregates (WB; 1995, p49-50, WB, 2005, p46-48), according to standard practice (Deaton and Zaidi, 2002, p23-38) :

- total food consumption includes purchased food, self-produced food, the estimated value of self-produced food and food received for free from friends and relatives;

- non-food consumption includes expenditures on clothing, utilities, services, personal care, hygiene items, communication, transportation and other expenses;

- housing expenses were not included in the consumption aggregates due to a low level of renting expenses in the sample and a high level of the effect of including housing expenses in consumption in comparison with households' owning dwellings;

- expenditures on durable goods are excluded from the consumption aggregates, but semi-durable goods are included, together with the calculation of the imputed user value for durable goods (depreciation rate).

However, the National Statistical Committee (NSC, 2012c, p1-4) introduced some additional changes in the calculation of consumption aggregates, based on the aggregate expenditures since 2002, affecting the poverty line (NSC), including expenses on purchase:

1 immovable property,

2 durable goods,

3 livestock, poultry and bees,

4 purchase of seeds and fertilizers,

5 expenses on gifts, assistance for relatives and friends,

6 production costs, including veterinarian services,

7 taxes, duties and other fees.

Later in the official reports were noticed that method based on personal consumption expenditures used for defining poverty rates in the country since 2004 (NSC, 2011c, p18). Aggregation indexes, Headcount, Poverty Gap and Poverty Severity Indexes, are typical for the poverty measurement in developing countries and no specific changes were introduced here.

The *questionnaire* used in the Living Standards Surveys is a typical tool suggested by LSMS methodology. It changed several times, however certain general characteristics have been kept to collect comprehensive high-quality data about household well-being (KMPS, 1994c, p1-51; KMPS, 1994d, p1-31; KMPS, 1994b, p1-50; KPMS, 2002a, p1-89; NSC, 2008, p1-114):

- the household questionnaire consists of two parts filled in during two-round visits with a two-week interval;

- the household questionnaire consists of:

    [1] demographic information,

    [2] housing and infrastructure,

    [3] household consumption, including self-production,

    [4] activities in agricultural and non-agricultural areas

    [5] personal data on health, education, migration, and labor information

- the Population Point Questionnaire, initially collected separately (KMPS, 1994e, p1-15), was later joined with the Price Questionnaire (KMPS, 1994a, p1-39), but with almost no changes to its content, which consists of the population and infrastructure features and price data of the selected PSU (KPMS, 2002b, p1-15; WB, 1995, p48; WB, 2005, p7,45-46).

The structure and content of the questionnaire is discussed additionally and in more detail further below in the part that explains the implemented survey tools and structure.

The methodology of the surveys changed several times within the history of living standards surveys in Kyrgyzstan. As we can see from Table 3.2, two types of surveys were introduced in the country. One type is the Living Standards Measurement Study Survey, introduced by

Table 3.2: List of the Poverty Monitoring Surveys in Kyrgyzstan

Surveys	Years	Sample Size	
		Households	Persons
Kyrgyz Multipurpose Poverty Survey (KMPS)	1993	1938	9066
Kyrgyz Poverty Monitoring Survey (KPMS)	1996	1951	8995
Kyrgyz Poverty Monitoring Survey (KPMS)	1997	2604	13633
Kyrgyz Poverty Monitoring Survey (KPMS)	1998	2979	15329
Household Energy Survey (HES)	1999	2994	4960
Household Budget Survey (HBS)	1999-2003	1081	4756
Kyrgyz Integrated Household Survey (KIHS)	2003-present	4760	19516

Source: WB poverty reports, 1995-2007

and adopted from the World Bank guidance by the Kyrgyz Republic's National Statistical Committee (NSC). Another type is the Household Budget Survey, carried out by the NSC and originating from the Soviet experience of measuring welfare, but only the true panel survey among all of the poverty surveys. Unfortunately it is not representative at the regional level (WB, 2005, p25). The Household Energy Survey (HES) was implemented under the financial and technical support of the DFID. Despite the fact that the HES was primarily designed for investigating energy consumption issues on the household level, "the NSC did, however, ensure that the survey design was such that the variables used for measuring poverty were the same as those in the earlier KMPS surveys" (WB, 2001, p2). The last modification of the surveys happened in 2003 with the Kyrgyz Integrated Household Survey (KIHS), which may be understood as a return to the LSMS methodology.

The KIHS was modified with the financial and technical support of the DFID. Oxford Policy Management developed and launched this survey with the NSC. In certain characteristics the KIHS differs from the other surveys which had been implemented in Kyrgyzstan at an earlier time (Esenaliev et al., 2011, p2):

- it has the biggest sample size (around 5000 households);

- the sample structure is a mixture between panel and cross-section type (maximum one-quarter of the sample changes annually);

- consumption data are collected quarterly.

Originally it was planned to use the KIHS database as the basis for an additional alternative test of the hypothesis, but certain problems with regard to data quality and the clarity of weighting procedures were detected by researchers (Esenaliev et al., 2011, p5). The situation concerning the database was complicated further by the absence of official documentation. To avoid these problems, a serious reworking of the database would have been required. Due to serious limits of the available resources, this task was deferred and will be kept in view for a future separate research.

Sample selection: From the sampling point of view, two methods were applied. LSMS methodology uses probability sampling, but the Household Budget Survey, the standard income and expenditure survey of the republics of the former Soviet Union, uses quota sampling and, thus, cannot be extrapolated to the national population (WB, 2007a, p18).

The LSMS sample methodology will be discussed only because currently HBS is not relevant. In 1993 a stratified, multi-stage sampling procedure was used to provide probability samples (Ackland, 1996, p13-18). On the basis of regional distribution, the country was divided into two types of strata - self-representing and non-self-representing units. Depending on the representativeness of the strata, different stages of the procedure were applied. The 12 self-representing PSUs included all the main urban territories in Kyrgyzstan, with a population of 34.2%, and 718 households were selected within the two-stage procedure from the planned 2100. The remaining amount of the sample was selected from the remaining 40 rayons, covering all the rural areas. The following strata was applied, depending on climate, agriculture and ethnic pattern:

1 mountains, agriculture and animal husbandry, primarily Kyrgyz population.

2 mountains, agriculture and animal husbandry, as well as nurseries, primarily Kyrgyz population.

3 mountains, agriculture and industry, Kyrgyz and Uzbek population.

4 valleys, agriculture, Kyrgyz mixed with Russian-speaking population.

5 valleys and mountains, agriculture, Kyrgyz mixed with Uzbek population.

6 valleys, agriculture and industry, Kyrgyz mixed with some Russian-speaking population.

The population of strata 2 and 5 was twice the size of the remaining ones, therefore each was split into two strata, so that there were finally eight strata. Thus, the multi-stage sampling procedure was implemented as follows:

1 In the 12 self-representing urban PSUs, a two-stage procedure was applied for:

[-] selecting census enumeration districts,

[-] selecting households,

2 In the urban areas of non-self-representing PSUs, a four-stage procedure was applied for:

[-] selecting two PSUs, if there were more than two units in a stratum,

[-] selecting the urban population point,

[-] selecting census enumeration districts,

[-] selecting households,

3 In rural areas of non-self-representing PSUs, a three-stage procedure was applied for:

[-] selecting two PSUs, if there were more than two units in a stratum,

[-] selecting the village,

[-] selecting the households.

It needs to be added that in 1993 the real base for the household was the list of dwellings for each of the selected PSUs, because no current national census data existed. In 1996, this information was updated through the Kyrgyz Household Registration System and in 1996-1998, the procedure of stratification was simplified. Seven strata defined by oblasts (see Map 2.2 on page 8) and the capital Bishkek were selected. In each of the strata, strata for the rural and urban population were selected, while in Bishkek only the urban population was accounted for. The different procedures used in urban and rural areas were:

1 In urban areas, a three-stage procedure was applied for:

[-] selecting large (with a population of more than 41,125 inhabitants) and small towns,

[-] selecting Census Posts,

[-] selecting the households,

2 In rural areas, a two-stage procedure was applied for:

[*] selecting Aiyl Aimak [1] and population points in rural areas,

[*] selecting the households from selected Census Posts and Aiyl Kenesh.

---

[1] Aiyl Aimak or Aiyl Okmotu is the smallest territorial subdivision in the Kyrgyzstan, is a part of a rayon; it is typically join in one group 2 or 3 villages, but sometimes consist of one village only

Originally, the samples for 1993 were developed by the WB research mission, the samplings for the 1996 and 1998 surveys were adopted by the specialists of the National Statistical Committee. In 2003, samples were stratified in a two-stage random sampling, using the 1999 population census with the support of Oxford Policy Management. 15 strata represent the urban and rural population of seven Oblasts and Bishkek. Each cross-section is representative at the national and regional level and at the urban and rural stratum too (Esenaliev et al., 2011, p2-5).

With this, the subsection on the main important features of the poverty measurement methodology and its application in Kyrgyzstan comes to an end. The poverty measurement methodology described in the Poverty Section, including poverty indexes, survey organization, sampling methodology, will further be used in the parts devoted to poverty survey planning and implementation in the selected target region of Kyrgyzstan, as well as in a proposed hypothesis validation process.

3.2   Production Frontier History

The current section describes the main stages of development and the pillars of the specific 'frontier' area of production efficiency literature, as well as the search for a potential methodology appropriate for the analysis of rural producers in the developing countries. The production frontier theory is a specific production function direction which began to develop separately from the mainstream of the production function field with the publication of the important work of Farrell (1957). Earlier works on the similar efficiency problem exist, written by Koopmans (1951) and Debreu (1951), but Farrell was the first to develop the idea right up to the point of empirical application. Today, this new direction is widely known as the frontier methodology (Coelli et al., 2005, p51).

The core advantage of the frontier approach is that in the traditional production function there exists a hidden contradiction in presuming maximum output from the input factors, while the estimation always concentrates on average production possibilities, widely observed in practice. Farrell (1957) proposed to compare the output of a particular firm (or any production unit) to evaluate how the inputs are used to produce the final output. This evaluation process is based on the estimation of a benchmark frontier in which the firm is assessed by analyzing the used inputs and final outputs. The level of efficiency of the specific firm is thus measured as the distance from the frontier, which is the maximum potential output that can possibly be reached with the invested inputs.

According to Farrell (1957), the production unit consist of two different components of efficiency: technical efficiency (TE), which demonstrates the capacity of a producer to obtain maximal output from a given set of inputs, and allocative efficiency (AE), which reflects the capacity of a firm to use the inputs in optimal ratios, taking into account their applicable prices and the technology level. These two measures in combination provide us with the unified measure of total economic efficiency.

Two main reasons play a significant role in the popularity of the concept and its further development in the production literature:

- the model is consistent with the major definitions of production microeconomics - production, profit and cost function (Forsund et al., 1980, p5-25),

- the model supports the possibility to estimate a specific efficiency measure for a specific firm (Lau and Yotopordos, 1971, p95).

Farrell's work became the basis for a significant amount of frontier type models. All of them may be divided into two different groups: parametric and non-parametric. Parametric frontiers are limited by a specific functional form (e.g., Cobb-Douglas), while non-parametric frontiers are free from a functional form, as well as from specific functional assumptions. Another important differentiation in frontier model classification is made between deterministic and stochastic frontiers. The deterministic approach proposes that any possible deviation from the frontier is caused by inefficiency, while the stochastic approach additionally allows for the possibility of statistical noise (Bravo-Ureta and Pinheiro, 1993, p89).

The non-parametric deterministic approach was introduced by Charnes et al. (1978). It provides the foundations for efficiency measurement, using linear programming methods labelled Data Envelopments Analysis (DEA) for the constant returns to scale case. Further modifications for variable returns to scale were presented by Banker, Charnes and Cooper in 1984 (Färe et al., 1994, p93).

The parametric deterministic approach was first introduced by Aigner and Chu (1968). They estimated a Cobb-Douglas production frontier through linear and quadratic programming techniques. This procedure was further developed by introducing the probabilistic frontier production model.

Another important method is the econometric method based on the Stochastic Frontier Analysis (SFA). The model incorporates a composed error structure with a two-sided symmetric and a one-sided component, developed independently by Aigner et al. (1976), and Meeusen and van den Broeck (1977). The one-sided component reflects the inefficiency effect, while the two-sided error captures the random effects outside the control of the production unit, including measurement errors and other statistical noise typical of empirical relationships. The estimation of a stochastic frontier function can be accomplished in two ways. First, if no explicit distributional assumption for the efficiency component is made, the production frontier can be estimated by a stochastic version of corrected ordinary least squares (COLS), suggested by Afrait (1972) and Richmond (1974). On the other hand, if an explicit distribution is assumed, such as exponential, half-normal or gamma, then the frontier is estimated by maximum likelihood methods. Maximum likelihood estimates (MLE) make use of the specific distributions of the disturbance term, and, are thus more efficient than COLS (Greene, 1980, p54-55). The initial inability of calculating an individual firm's efficiency measures from the stochastic frontier model was overcome by the work of Jondrow, Lovell, Materov and Schmidt as reviewed by Kumbhakar and Lovell (2000, p9).

In the next subsections we analyze the main properties of the frontier model assumptions and two main production frontier methods - Data Envelopment Analysis (DEA) and Stochastic Frontier Analysis (SFA).

3.2.1 Frontier Model Assumptions

In the settings of the estimation of maximum potential boundaries, which envelop data, in contrast with the functions, which intersect data, the main purpose of productivity analysis studies, is to evaluate the performance of a firm from the perspective of technical efficiency. The problem to be analyzed is thus set in terms of physical input and output quantities. We propose that data should be available in cross-sectional form to evaluate the inputs and outputs used in the production process. Measuring efficiency for this data set first of all requires a definition of the production sets boundary; then the distance between any observed point and the boundary of the production set needs to be measured (Daraio and Simar, 2007, p20).

The following production frontier model assumptions are based on the explanation by Daraio and Simar(2007, p20-21). *The production set*, Ψ, with the list of p inputs and q outputs can be defined as follows in the Euclidean space $\Re_+^{p+q}$:

$$\Psi = \left\{ (x,y) | x \in \Re_+^p, y \in \Re_+^q, (x,y) \text{ is affordable} \right\}, \qquad (3.4.)$$

where x is the input variables vector, y is the output variables vector and "affordability" of the vector (x,y) means that within the firm it is possible to produce the output product $y_1, \ldots, y_q$, when the input quantities $x_1, \ldots, x_p$, are invested (all quantities counted per unit of time).

The *input requirement set* (for all $y \in \Psi$) can be specified as:

$$C(y) = \left\{ x \in \Re_+^p | (x, y) \in \Psi \right\}. \tag{3.5.}$$

The input set $C(y)$ consist of all input vectors that can yield the product vector $y \in \Re_+^q$. The *output correspondence set* (for all $x \in \Psi$) is defined as:

$$P(x) = \left\{ y \in \Re_+^q | (x, y) \in \Psi \right\}. \tag{3.6.}$$

$P(x)$ consist of all output product vectors that can be yielded by a given input vector $x \in \Re_+^p$.

The production set Ψ can be reconstituted from the proposed inputs set:

$$\Psi = \left\{ (x, y) | x \in C(y), y \in \Re_+^q \right\}. \tag{3.7.}$$

Moreover, it considers that:

$$(x, y) \in \Psi \Leftrightarrow x \in C(y), y \in P(x\}, \tag{3.8.}$$

which shows that the product and input sets are analyzed as equivalent representations of the technology, as both of sets are belongs to Ψ

The efficient boundaries of the sections of Ψ can be defined in radial terms[II] as follows in the input and product dimensions.

In the input space in the form:

$$\partial C(y) = \left\{ x | x \in C(y), \theta x \notin C(y), \forall \theta, 0 < \theta < 1 \right\} \tag{3.9.}$$

and in the output product space in the form:

$$\partial P(x) = \left\{ y | y \in P(x), \lambda y \notin P(x), \forall \lambda > 1 \right\} \tag{3.10.}$$

For the input space notation this means that the corresponding decrease of input does not directly lead to a subsequent decrease of output and is defined by the value of θ, which is independent from the output vector. At the same time, it means for the output boundary that an increase of the output may be proportional or disproportional to the increase of the input, depending on the λ value, which is independent from the input vector. More detailed descriptions of the axioms of productions is presented by Färe and Grosskopf (2004, p151-161).

[II] The idea suggested by Farrell to measure a radial distance from the point of firm produce output to its corresponding frontier.

3.2.2 Data Envelopment Analysis

The following DEA model construction is exposed from the summary explanation described in Daraio and Simar (2007, p31-33). DEA estimated efficiency assumes free distribution and a convex character of the production set Ψ. The estimation includes efficiency assessment for a given unit *(x, y)* relative to the borders of the limits of $X = \{ (X_i, Y_i), \text{ where } i = 1, \dots, n$ are the observation units$\}$:

$$\hat{\Psi}_{DEA} = \left\{ (x,y) \in \Re_+^{p+q} | y \leq \sum_{i=1}^{n} \gamma_i Y_i; x \geq \sum_{i=1}^{n} \gamma_i X_i, \text{for } (\gamma_1, \dots, \gamma_n) \right.$$
$$\left. \text{s.t.} \sum_{i=1}^{n} \gamma_i = 1; \gamma_i \geq 0, i = 1, \dots, n \right\} \tag{3.11.}$$

Thus $\hat{\Psi}_{DEA}$ is the smallest free distributed production output set covering all the data. The $\hat{\Psi}_{DEA}$ in 3.11. presumes Variable Returns to Scale (VRS) and is typically labelled as $\hat{\Psi}_{DEA-VRS}$. It may be transformed to other types of scale options. It is possible for:

- *Constant Returns to Scale* (CRS) if the equality requirement constraint $\sum_{i=1}^{n} \gamma_i = 1$ in 3.11. is excluded;

- *Non-Increasing Returns to Scale* (NIRS) if the equality requirement constraint $\sum_{i=1}^{n} \gamma_i = 1$ in 3.11. is changed to the form $\sum_{i=1}^{n} \gamma_i \leq 1$;

- *Non-Decreasing Returns to Scale* (NDRS) if the equality requirement constraint $\sum_{i=1}^{n} \gamma_i = 1$ in 3.11. is modified to the form $\sum_{i=1}^{n} \gamma_i \geq 1$.

The evaluation of an input set is given for all y by: $\hat{C}(y) = \{x \in \Re_+^p | (x,y) \in \hat{\Psi}_{DEA}\}$ and $\partial C(y)$ defines the input frontier limits for y.

For a firm working at level (x_0, y_0) the assessment of the input efficiency score $\theta(x_0, y_0)$ is received by solving the following linear model (considering the VRS situation here):

$$\hat{\theta}_{DEA}(x_0, y_0) = \inf \left\{ \theta | (\theta_{x_0, y_0}) \in \hat{\Psi}_{DEA} \right\} \tag{3.12.}$$

$$\hat{\theta}_{DEA}(x_0, y_0) = \min \left\{ \theta | y_0 \leq \sum_{i=1}^{n} \gamma_i Y_i; \theta_{x_0} \geq \sum_{i=1}^{n} \gamma_i X_i; \theta > 0; \right.$$
$$\left. \sum_{i=1}^{n} \gamma_i = 1; \gamma_i \geq 0; i = 1, \dots, n \right\}. \tag{3.13.}$$

$\theta(x_0, y_0)$ defines the radial distance between (x_0, y_0) and $(\hat{x}^\theta(x_0|y_0), y_0)$ where $\hat{x}^\theta(x_0|y_0)$ is the sum of the inputs the unit should apply in order to be on the "efficient frontier" of $\hat{\Psi}_{DEA}$ with the same level of final product, y_0, and the same ratio of inputs; i.e. moving from x_0 to $\hat{x}^\theta(x_0|y_0)$ along the isoquant curve $\theta(x_0)$. The trace of x_0 on the efficient frontier is thus equivalent to $\hat{x}^\theta(x_0|y_0) = \hat{\theta}(x_0, y_0)x_0$

For the output-oriented case, the same procedure is repeated with the necessary changes. The output product set is assessed by: $P^{(}x) = \{y \in \Re_+^q | (x, y) \in \hat{\theta}_{DEA}$, and $\partial P^{(}x)$ output frontier limits for x.

The assessment of the output product efficiency ratio for a given (x_0, y_0) is received by solving the following linear model:

$$\hat{\lambda}_{DEA}(x_0, y_0) = \sup\{\lambda | (x_0, \lambda_{y_0}) \in \hat{\Psi}_{DEA}\} \qquad (3.14.)$$

$$\hat{\lambda}_{DEA}(x_0, y_0) = \max\{\lambda | \lambda y_0 \leq \sum_{i=1}^{n} \gamma_i Y_i; x_0 \geq \sum_{i=1}^{n} \gamma_i X_i; \lambda > 0;$$
$$\sum_{i=1}^{n} \gamma_i = 1; \gamma_i \geq 0; i = 1, \dots, n\}. \qquad (3.15.)$$

In Fig. 3.1 the graphical illustration of the DEA estimator is presented. Additionally, the *slacks* concept is also illustrated. On the left graph describing input-oriented DEA, we suppose that all farms produce the same amount of product, but the inputs are different. It is evident that farm A could produce the same unit y with the lower input level of x_1. Thus moving from point A to point B could decrease inputs from 3 to 2. This situation is specified as *input slack*. Despite the fact that the firm is efficient, there is a *surplus* of input x_1. We may conclude that there is a slack in input $j = A - B$ of the firm i, i.e., x_i^j, if the following condition,

$$\sum_{i=1}^{n} \gamma_i x_i < x^j{}_i \hat{\theta}(x_i, j_i), \qquad (3.16.)$$

is true for some solution with the value of $\lambda_i, i = 1, \dots, n$ (Daraio and Simar, 2007, p32). A similar scheme is applied for the output-oriented case, i.e., the case of firm L, which could increase the production of y_i by moving from L to M. The right part of Fig. 3.1 demonstrates this case.

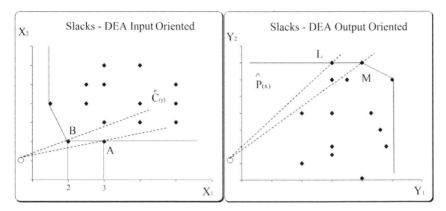

Figure 3.1: Input and Output Slacks
Source: Author's own representation following Färe et al. (1994)

3.2.3 Stochastic Frontier Analysis

Another important frontier approach, known as Stochastic Frontier Approach (SFA), was proposed independently by Aigner et al.(1976) and Meeusen et al. (1977) (Kumbhakar and Lovell, 2000, p8). They proposed an idea of the estimation of a function by specification of the frontier error term, which is itself admitted to be a stochastic parameter which, from the production function for the ith firm may be expressed in the following form:

$$\ln y_i = \acute{x}_i \beta + \nu_i - u_i = \acute{x}_i \beta + \varepsilon_i, \qquad (3.17.)$$

where,

- y_i is the production output of ith firm,

- x_i is the $K \times 1$ vector of the logarithm of inputs of the ith firm,

- β is the unknown parameters vector,

- ν_i is the two-sided "statistical" noise component, which is assumed to be independently and identically distributed (iid) and symmetrically distributed independently of u_i,

- u_i is the nonnegative technical inefficiency component of the error term.

Thus, through introducing a new component ν_i, the error term $\varepsilon = \nu_i - u_i$ became non-symmetric, because u_i is restricted to being positive.

Model 3.17., is called a *stochastic* frontier production, because the output range is bounded from above by the random variable $exp(\acute{x}\beta + \nu_i)$. ν_i can be positive or negative and the final

39

range of the stochastic frontier product is specified by the deterministic part of the model, $exp(\dot{x}\beta)$. The visual presentation of the model is possible if we narrow down the model to the *one input* x_i - *one output* y_i model type. Finally model 3.17. is defined as follows:

$$y_i = q_i = \underbrace{\exp(\beta_0 + \beta_1 ln x_i)}_{\text{deterministic component}} \times \underbrace{\exp(\nu_i)}_{\text{noise}} \times \underbrace{\exp(-u_i)}_{\text{inefficiency}}, \qquad (3.18.)$$

In Fig. 3.2, a frontier is determined for the inputs and products of two production units, farms A and B, with an applied deterministic frontier model reflecting the diminishing returns to scale. The input area is located along the horizontal vector of x_i. Product values are arranged in the limits of the vertical vector y_i. Farm A employs the input X_A to make product q_A. Farm B utilizes the input X_B to make the output q_B (two observations indicated by the points marked with ▲). If there were no inefficiency effects (i.e., if $u_A = 0$ and $u_B = 0$), the so-called *frontier* outputs would be equal:

$$q_A^* = \exp(\beta_0 + \beta_1 ln x_A + \nu_A),$$

and

$$q_B^* = \exp(\beta_0 + \beta_1 ln x_B + \nu_B),$$

for farms A and B accordingly. The values of the frontiers are indicated by the points marked with ⊗. Farm A's frontier output lies visibly above the settled part of the production frontier, simply because it has a *positive* noise effect part (i.e., $\nu_A > 0$), while for Farm B the frontier range is located under the deterministic component of the frontier, due to a contrarily demonstrated negative value of the error term (i.e., $\nu_B < 0$).

Another effect is seen for the observed product of Farm A which is located below the settled component part of the frontier because the joined values of the inefficiency and noise components are negative (i.e., $\nu_A - u_A < 0$).

The basic idea of the stochastic frontier approach is aimed at predicting the individual firm's (production unit's) inefficiency effects. If we want to search for output-oriented measures of technical inefficiency, the main inefficiency indicator is to be found through defining the ratio of the observed output in correlation to the proposed stochastic frontier product, which may be formulated as follows:

$$TE_i = \frac{q_i}{\exp(x_i'\beta + \nu_i)} = \frac{\exp(x_i'\beta + \nu_i - u_i)}{\exp(x_i'\beta + \nu_i)} = \exp(-u_i) \qquad (3.19.)$$

The calculated technical efficiency index lies between zero and one. The index evaluates the product of the ith production unit relative to the potential product that a fully-efficient producer exploiting the same input factors may produce. As mentioned earlier ν_i is distributed

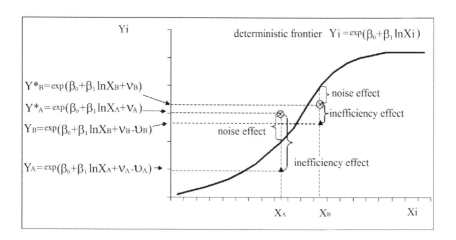

Figure 3.2: Stochastic Frontier Scheme
Source: Author's own representation following Coelli et al. (2005)

independently of u_i, and both error components are uncorrelated with the explanatory input values in $\mathbf{x}_i$. The other assumptions are relevant for the model 3.17. with two random terms, a symmetric error ν_i and a non-negative random variable u_i, as follows:

$$E(\nu_i) = 0, \qquad \text{(zero mean)} \qquad (3.20.)$$
$$E(\nu_i^2) = \sigma_\nu^2, \qquad \text{(homoscedastic)} \qquad (3.21.)$$
$$E(\nu_i\nu_j) = 0 \quad \text{for all} \quad i \neq j, \qquad \text{(uncorrelated)} \qquad (3.22.)$$
$$E(u_i^2) = \text{constant}, \qquad \text{(homoscedastic)} \qquad (3.23.)$$
$$E(u_iu_j) = 0 \quad \text{for all} \quad i \neq j, \qquad \text{(uncorrelated)} \qquad (3.24.)$$

As can be seen from the assumptions ν_i, the noise component has the same assumptions as in the classical linear regression model. The same could be concluded with regard to the inefficiency coefficient, except it has a non-zero mean (as $u_i \geq 0$).

Based on the proposed assumptions, the estimators of the *slope* coefficients is estimated the maximum likelihood (ML) method. Other methods like COLS, is less appropriate as specified earlier on page 35.

The following distributional assumptions are derived from the explanation described in Coelli et al. (2005, p245-247); Kumbhakar and Lovell (2000, p69,74-80).

The *Normal-Half Normal Stochastic Frontier Model Specification* (Aigner et al., 1976, p24) proposed in model 3.17. suggests the following assumptions for this type of distribution,

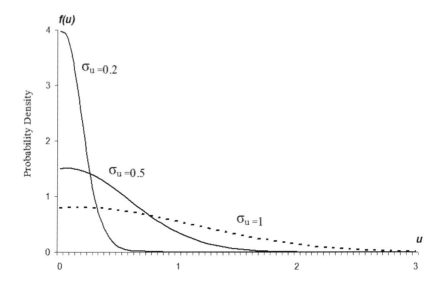

Figure 3.3: Half Normal Distributions
Source: Author's own representation following Kumbhakar and Lovell (2000)

which is the basic specification of that type:

$$\nu_i \sim iidN(0, \sigma_\nu^2), \tag{3.25.}$$

$$u_i \sim iidN^+(0, \sigma_u^2) \tag{3.26.}$$

Assumption 3.25. defines that ν_i constitutes normal random variables in the model, which are independently and identically distributed, with zero mean and variances σ_ν^2. Assumption 3.26. proposes that u_i values are normal-half normal random variables, which are independently and identically distributed, with a scale parameter σ_u^2. The proposed probability density function (pdf) of each u_i is a normal random variable, in truncated type (only ≥ 0) with a zero mean and variance σ_u^2. Both of the variables are distributed independently from each other and the regressors too.

The distribution of the density function $u \geq 0$ for typical values is presented in Fig. 3.3. The function is specified as follows:

$$f(u) = \frac{2}{\sqrt{2\pi}\sigma_u} \exp\left\{-\frac{u^2}{2\sigma^2_u}\right\}. \tag{3.27.}$$

The density function of ν is

$$f(\nu) = \frac{1}{\sqrt{2\pi}\sigma_\nu}\exp\left\{-\frac{\nu^2}{2\sigma^2_\nu}\right\}. \tag{3.28.}$$

As we assumed that u and ν are independent from each other, the joint density function could be presented as the product of their individual density functions:

$$f(u,\nu) = \frac{2}{2\pi\sigma_u\sigma_\nu}\exp\left\{-\frac{u^2}{2\sigma^2_u} - \frac{\nu^2}{2\sigma^2_\nu}\right\}. \tag{3.29.}$$

As $\varepsilon = \nu - u$, the joint density function for u and ε is

$$f(u,\varepsilon) = \frac{2}{2\pi\sigma_u\sigma_\nu}\exp\left\{-\frac{u^2}{2\sigma^2_u} - \frac{(\varepsilon+u)^2}{2\sigma^2_\nu}\right\}. \tag{3.30.}$$

The marginal density function of ε is obtained by integrating u out of $f(u,\varepsilon)$. That brings us to

$$\begin{aligned}
f(\varepsilon) &= \int_0^\infty f(u,\varepsilon)du \\
&= \frac{2}{\sqrt{2\pi}\sigma}\left[1 - \Phi\left(\frac{\varepsilon\lambda}{\sigma}\right)\right]\exp\left\{-\frac{\varepsilon^2}{2\sigma^2}\right\} \\
&= \frac{2}{\sigma}\phi\left(\frac{\varepsilon}{\sigma}\right)\Phi\left(-\frac{\varepsilon\lambda}{\sigma}\right),
\end{aligned} \tag{3.31.}$$

where $\sigma = (\sigma_u^2 + \sigma_{2\nu})^{\frac{1}{2}}$, $\lambda = \frac{\sigma_u}{\sigma_\nu}$ and $\Phi(\cdot)$ and $\phi(\cdot)$ are the standard normal cumulative distribution and density functions. The normal-half normal distributions contain two basic parts, σ_u and σ_ν. The graphical presentation of the typical distributions of different combinations of σ_u, and σ_ν is seen in Fig. 3.4. As distributions are asymmetric due to $\sigma_u > 0$, distributions are logically negatively skewed with the modes and means below zero values in each of the presented cases.

The marginal density function $f(\varepsilon)$ is asymmetrically distributed with mean and variance

$$\begin{aligned}
E(\varepsilon) &= -E(u) = -\sigma_u\sqrt{\frac{2}{\pi}} \\
V(\varepsilon) &= \frac{\pi-2}{\pi}\sigma_u^2 + \sigma_\nu^2
\end{aligned} \tag{3.32.}$$

Initially Aigner et al. (1976) suggested the use of mean technical efficiency of all producers $[1 - E(u)]$ as estimator. Later, Lee and Tyler (1978) according to Kumbhakar and Lovell (2000, p770) corrected the idea further by introducing:

$$E(exp-u) = 2[1 - \Phi(\sigma_u)]exp\left\{\frac{\sigma_u^2}{2}\right\}. \tag{3.33.}$$

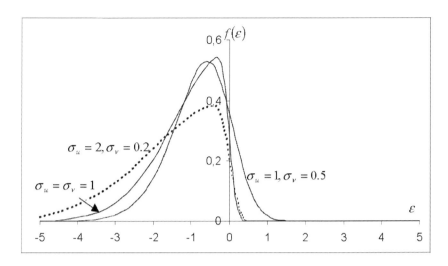

Figure 3.4: Normal-Half Normal Distributions
Source: Author's own representation of Kumbhakar and Lovell (2000)

This form is easier in calculations as $[1 - u]$ includes only the first term in the power series expansion of $\exp{-u}$. The log likelihood function for a sample of I producers is derived from 3.31.

$$\ln L = \text{constant} - I \ln \sigma + \sum_{i} \ln \Phi \left(-\frac{\varepsilon_i \lambda}{\sigma} \right) - \frac{1}{2\sigma^2} \sum_{i} \varepsilon_i^2. \qquad (3.34.)$$

This function could be maximized with respect to the parameters to receive maximum likelihood estimates of all parameters. Estimates are consistent for $I \rightarrow +\infty$.

The last step is to obtain estimates of the technical efficiency of each producer. We have estimates of ε, which obviously contain information on u. We suppose that there are not that many fully efficient producers in the sample. The same is proposed by the selected distributional assumption. The task is to extract the information on u. Different techniques have been developed since the definition of the stochastic frontier. Battese and Coelli (1988, p390) proposed an alternative approach of estimations:

$$
\begin{aligned}
TE_i &= E(\exp{-u_i}|\varepsilon_i) \\
&= \left[\frac{1 - \Phi(\sigma_* - \frac{\mu_{*i}}{\sigma_*})}{1 - \Phi(-\frac{\mu_{*i}}{\sigma_*})} \right] \cdot \exp\left\{ -\mu_{*i} + \frac{1}{2}\sigma_*^2 \right\}
\end{aligned}
\qquad (3.35.)
$$

The estimator given in 3.35. is more preferred, analogous to 3.33., especially in the cases when u_i is not close to zero values. Nevertheless, we need to specify additionally that the estimates of technical efficiency cannot be firmly consistent from an econometric point of view because the variation associated with the distribution of $(u_i|\varepsilon_i)$ is independent of i. Unfortunately it is the best level of accuracy that can be achieved with cross-sectional data (Kumbhakar and Lovell, 2000, p78).

Furthermore, production frontier researchers developed a number of distributional assumptions relating to the distribution of the combined random error, including the:

- Normal-Gamma Model,

- Normal-Exponential Model,

- Normal-Truncated Normal Model

However, the analysis shows that mean efficiencies are quite sensitive to the distributional assumptions. Researchers Ritter and Simar (1997) according to Kumbhakar and Lovell (2000, p90) suggest to use more simple distributional forms (Half-Normal or Exponential), which is more preferable due to lower sensitivity.

3.2.4 Frontier Model Application

We analyze the peculiarities of the frontier methodology concerning the definition of technical efficiency for individual producers. In this review, we concentrate mainly on the cross-section models and appropriate assumptions, as we are able to obtain them. SFA developments with panel data demonstrate better results, but we must skip them for the future research. Two types of methods are the most popular among the frontier methods: Data Envelopment Analysis and Stochastic Frontier Analysis. DEA involves mathematical programming tools, and SFA is based on econometric methods. Each of the approaches has certain peculiarities and problems.

The DEA's main problem is its inability to distinguish the source of the statistical noise and the following measurement error. All deviations from the frontier are supposed to result from technical inefficiency. The DEA approach is deterministic and non-parametric, and it is better from the methodological point of view because it does not require a specific functional form.

SFA allows for random disturbances, such as fluctuations of the weather conditions, the effects of pest invasion and possibilities of diseases, and also for the measurement errors in the output variables. Therefore, the SFA approach is chosen as more appropriate for agricultural applications, especially in developing countries (Coelli et al., 2005, p202, 305). However, a negative consequence of the SFA methods is the strong influence of the distributional assumptions of error term.

Comparing both methods with regard to their application to the problem of determining the efficiency of rural households, we decided to use both methods. In theory, the difference in results between DEA and SFA may bring about new information about the methods' applicability and interpretation of the results in empirical analysis.

4. Survey

This chapter is dedicated to describing the goals, methods and content of the field study implemented in fall 2010 and in spring-summer 2011 in Talas Oblast, Kyrgyz Republic, and to describing the features of the target zone. The field study was implemented in two stages - the pilot (test) micro-survey in October-November 2010 and the full field study organized in June-May 2011. The tasks for the pilot study were to test and define the field study methodology, check and modify the questionnaire according to the study goals and available resources in the few selected villages, collect information about the target territory, prepare information support and check the local personal potential. The full field study was organized to implement the survey across the whole region according to the sample method in all four rayons (regional subdivisions) of Talas Oblast.

The chapter consists of three consecutive sections, which describe the survey stages. In the first section we provide the background of the territory, the following section includes the selected methodology, as well as the process of defining the questionnaire and sampling methodology. The next section provides the survey organization process in more detail, including training staff in field description and collected data quality control, and the final result of the survey.

4.1 Selected Region Description – Talas Oblast

4.1.1 Geography, Climate, Population and Economy

The target region was selected for a detailed research study. Originally it was suggested to study the most remote and undeveloped region in Kyrgyzstan - Batken Oblast, but due to the Kyrgyz-Uzbek ethnic conflict in June 2010, chances to fully and safely cover the Uzbek part of the population seriously decreased. State control in the area was low and the situation was quite unstable at the time of planning the survey. Therefore it was decided to change the target region. Finally, Talas Oblast was selected.

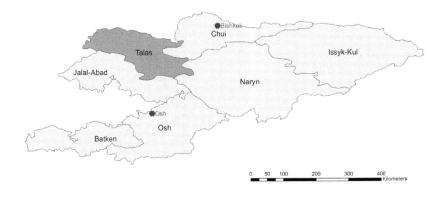

Figure 4.1: Location of the Target Region - Talas Oblast
Source: Modified from http://reliefweb.int/maps

Talas Oblast was singled out because of several factors. The main reasons for selecting the region was a constantly higher poverty level in comparison with the national level, remoteness from the main markets, prevalence of the rural sector, and high ethnic homogeneity.

Talas Oblast is one of the most remote areas of Kyrgyzstan. The Oblast is located in the north-western part of the country. In the north it borders on Kazakhstan, the border with Uzbekistan lies on a high mountain range.

The Oblast is connected with other parts of the country through high-pass mountain roads. Its connection with the capital through Kazakh territory is difficult due to political reasons, and the situation has worsened even more after the Kyrgyz political crisis in 2010 (TSO, 2011, p2).

Administratively, the region is divided into four districts (rayons): Talas , Bakai-Ata, Kara-Buura and Manas (see Fig. 4.2 on the page 49). The population consist of almost 220 thousand people, most of which are rural inhabitants (84.7%). There is one regional center - the town of Talas - and one urban type village only. The rural population lives in 90 villages. There are several villages with a population of 5000 to 7000 inhabitants, the biggest village is home to more than 10 thousand people. Most of the villages, however, are relatively small and their population varies from 1000 to 3000 inhabitants.

Ethnically, Kyrgyz dominate the oblast (92%), and there are some minor communities of Russian, Kurdish and Kazakh people (NSC, 2010a, p35-36). In the Soviet era, a certain number of native Germans, who were relocated from Russia during the Second World War, lived in the oblast, but after the start of transition the majority of them returned to Germany.

The Talas area is a predominantly agrarian region, with agriculture accounting for nearly two thirds of the area's product (64.3% in regional GDP in 2008); other main sectors are construction (9.1%) and trade (8.6%). The construction activity is mainly explained by investment in the mining sector (gold), which began in 2008. Unfortunately, the political collisions of the last years have more or less put the investment process in the mining sector on ice. Some activity is also provided by real estate services, the state and education (share of each sector approximately 3%). Other activity is relatively small and has not strongly affected the development of the region (TSO, 2011, p3-4).

In agriculture, the main cultures are haricot beans (38% of arable land), grain cultures (16%) and potatoes (15%). The main livestock are cattle (65 thousand heads) and sheep (400 thousands sheep). There are approximately 35 thousand rural households with an average size of more then 5 persons per household. The irrigation system is in relatively good condition. The average land share exceeds 2 hectares of land per household (TSO, 2011, p10-11).

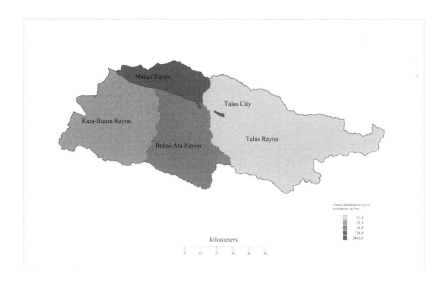

Figure 4.2: Talas Oblast - Rayons with Population Density
Source: modified from map of Talas Oblast Census 2009, NSC

4.1.2 Agriculture of the Region

Talas Oblast is characterized by a stagnating livestock sector and a more dynamic crop production. In the crop production, around half of the land is occupied by haricot beans, while the next two biggest cultures are wheat and potatoes. Other cultures are not significant in the overall picture. Another important culture, which is not included in official statistics, is the production of hay for the feeding of livestock.

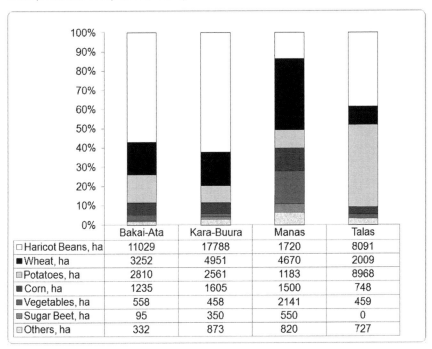

	Bakai-Ata	Kara-Buura	Manas	Talas
☐ Haricot Beans, ha	11029	17788	1720	8091
■ Wheat, ha	3252	4951	4670	2009
▨ Potatoes, ha	2810	2561	1183	8968
■ Corn, ha	1235	1605	1500	748
▨ Vegetables, ha	558	458	2141	459
▨ Sugar Beet, ha	95	350	550	0
☐ Others, ha	332	873	820	727

Figure 4.3: Crop Production by Districts in Talas Oblast in 2010 (ha)
Source: Talas Oblast Statistical Office, 2010

The haricot beans' prevalence is a relatively new trend in the region, which started in 2000. The culture became important for the region as a result of the efforts of Turkish merchandisers to introduce the culture in cooperation with the local authorities, and provides a stable export market. Since 2005, the culture has started to dominate the region and push aside the formerly leading wheat and sugar beat production. The second most important growing culture in the region is potatoes.

The regional pattern of crop structure varies depending on location. The more mountainous Talas Rayon basically concentrates on potatoes and beans in equal proportion. Bakai-Ata

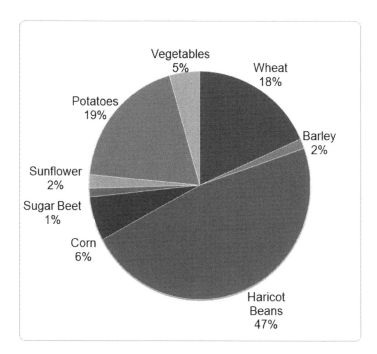

Figure 4.4: Crop Production Structure in Talas Oblast in 2010 (ha)
Source: Talas Oblast Statistical Office, 2010

and Kara-Buura Rayons, occupying the middle Talas Valley, are characterized by an extreme prevalence of haricot beans (around 60% of the crop). A more or less diversified crop structure is observed in the lowest part of the oblast, Manas Rayon. A hotter climate and close proximity to the Kazakh border and market stimulates the production of vegetables and sugar beet. More wheat and corn for livestock production were produced here.

The average productivity of cultures is not high in comparison with the average productivity in the developed countries. The main reasons for the low productivity are a weak investment potential of the sector, backward technology, undeveloped market logistics and an absence of infrastructure services.

Over the last years, the livestock sector has demonstrated low growth rates. This stagnation was defined by the factor of scarcity of winter pastures in most of the territory's regions and by the necessity to invest resources in forage for animals in the winter period. Survey observations show that at household level farmers prefer to specialize mainly on the crop sector. Due to lower investment capital and labor inputs, most farmers concentrate on crop production with some livestock production for their own subsistence consumption (cattle

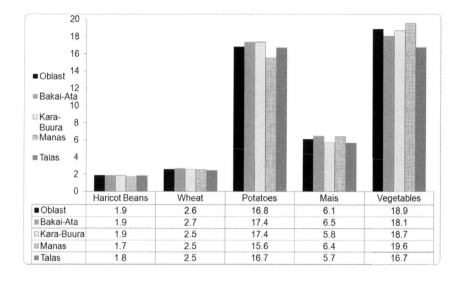

	Haricot Beans	Wheat	Potatoes	Mais	Vegetables
■ Oblast	1.9	2.6	16.8	6.1	18.9
■ Bakai-Ata	1.9	2.7	17.4	6.5	18.1
◻ Kara-Buura	1.9	2.5	17.4	5.8	18.7
▥ Manas	1.7	2.5	15.6	6.4	19.6
■ Talas	1.8	2.5	16.7	5.7	16.7

Figure 4.5: Crop Productivity in Talas Oblast in 2010 (ton per ha)
Source: Talas Oblast Statistical Office, 2010

and sometimes sheep). Livestock farmers mainly decide to grow sheep and horses.

In the summer, peasants with the subsistence number of livestock send their livestock to the remote summer mountain pastures with specialized livestock farmers, i.e. shepherds. Payment for herding services is calculated on a monthly basis, based on a price per head, depending on the type of animal cared for. Usually, the summer grazing period runs from May to October. In the winter time, livestock is grazed on local pastures near the villages. Typically, the situation is characterized by undergrazing of remote summer pastures and overgrazing of the nearest pastures. The system of collective grazing existed in the pre-transition period, but has disappeared today. New public institutional arrangements are in the process of being formed. The first arrangement was not effective and therefore a new system of pasture management has lately been introduced in the country, aiming at an optimization of pasture exploitation in the interest of the rural population. The pasture reform process is still in the initial implementation stage.

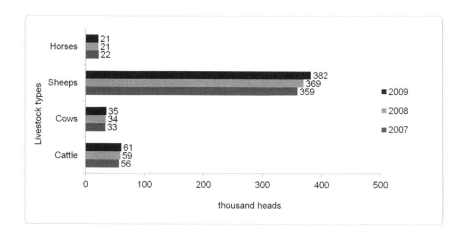

Figure 4.6: Livestock Dynamics in Talas Oblast, 2007-2009
Source: Talas Oblast Statistical Officce, 2010

4.2 Data Collection Methodology

The implemented survey aimed at collecting information on the poverty level in the selected
rural area of Talas Oblast as well as at gathering information on agricultural production issues
at the household level. It was based on the Living Standards Measurement Study (LSMS)
household survey methodology (Chander et al., 1980, p1-3). In the 1980s, LSMS was
specifically developed for less developed countries by the World Bank's poverty economists
and has since been conducted in about forty different developing countries. The main
purpose of these surveys was to collect comprehensive datasets of information at different
levels to correctly assess the poverty indicators and to evaluate the national government
policy effect on poverty (Abdul Wahab, 1980, p10). In Kyrgyzstan, five rounds of such
household surveys were implemented in the 90s (Temesgen et al., 2002, p1). They were
used by the local National Statistical Committee (NSC) of the Kyrgyz Republic as the basis
for creating domestic institutional arrangements of the living standards measurement under
the technical and financial support of the donor community in Kyrgyzstan, primarily by the
World Bank (WB). Currently, the living standards measurement survey is implemented on a
regular basis by the NSC and accepted by the WB as appropriate with regard to accuracy
and quality. LSMS methodology includes several key components - questionnaires, survey
organization procedure, and survey sampling procedures. Adopted and revised components of
the methodology (questionnaire, organization and sample) used in the project are described
in this chapter (Howes and Lanjouw, 1997, p3-8).

4.2.1   Survey Questionnaire

In Kyrgyzstan, the LSMS survey was carried out using a two-level questionnaire structure - a household questionnaire (KPMS, 2002a, p1-89) and a community (population point) questionnaire (KPMS, 2002b, p1-15). The household questionnaire was arranged to collect information on households (HH) - personal information of the members, consumption data, including their own production and exchanges in natural form. The community questionnaire was dedicated to the collection of price data and infrastructure features of population points.

During the research process, the survey questionnaires were seriously revised and restructured according to the primary research tasks and available resources (Ainsworth and van der Gaag, 1988, p9-10). Two important tasks defined the final version of the questionnaires. First of all, the survey was dedicated to the collection of detailed information on consumption and incomes of the rural HH in the area, to be able to define poverty indexes for each of the covered HH. In addition, the information was to cover production features information, including all the monetary costs, exchanges and additional incomes and costs of all the family members, in order to define the level of production efficiency of the selected households. Finally, it was also important to assess the process of the survey from an organizational point of view to balance available resources in the defined time. Due to the mentioned limitations, the questionnaires were seriously shortened and revised, but the main components containing core information were saved.

Another difference was the change of the questionnaire structure. It became a one level tool, covering household levels only. The main reason for excluding the population point questionnaire from our study was the different character of the research. The LSMS survey was developed for poverty measurement at a national level, while our research covered only one target area. Moreover, LSMS includes rural and urban areas, while our survey was limited to the rural population only. LSMS also supports the measurement of the monetary and non-monetary parameters of poverty, which requires the collection of additional qualitative information included in the community questionnaire. In our case, this information was not part of the research interest.

During the pilot stage, the necessity of measuring infrastructure differences and price information was analyzed. It defined two regional centers (markets), where all the prices for Talas Oblast were determined. However, the price difference between them at one moment in time was insignificant and primarily defined by transport costs as clarified by retail merchandisers. At the village level, price difference was not volatile. Thus the community questionnaire was skipped in our research.

The LSMS questionnaire was an extensive tool which required significant time costs. It required at least three to four hours to fill in all the parts of the questionnaire. The interview was divided into two parts, with a two week time difference. The first part collected all the general information on the household and the second part concentrated on rural activity, food production and consumption and was partly filled in by the HH members with information concerning food consumption. For some sections, all adult members had to be interviewed individually; additionally, it included a separate section concerning female health issues and anthropometric measurements (for children of 5 years and younger).

The given tool was modified by the NSC with the WB poverty specialists' support, because of the changes in methodology. The new revised version of the NSC questionnaire was split into three parts. General information collected on an annual basis from selected target HH, non-food commodities consumption collected on a quarterly basis. Food consumption data, as well as typical services and self-produced food data were collected on the basis of the diary filled in during the first two weeks of each quarter. The first two parts were filled in by the interviewer during the visit to the HH, while the last part was based on the diary completed independently by the responsible HH member and collected by on a quarterly basis by the interviewer. The original LSMS questionnaire and revised NSC questionnaires were used as initial basis to form our own specific research questionnaire. I redesigned it to fit the following requirements:

- the questionnaire should be filled in during one visit,

- the total length of the interview should be limited to 1.5 (maximum 2) hours,

- specific questions on the quality of land, infrastructure were included,

- all specific questions typical for urban areas were excluded,

- income questions were included in different sections to decrease suspiciousness toward the interview.

The questionnaire contains 11 major sections, covering different aspects of household activity. The sections are described in Table 4.1. More detailed information on the sections is given later, with comments on the changes done to optimize the content. The final version of the questionnaire is attached in Appendix A for details.

The **Household Card** section collects basic demographic information such as the number of residents, name, age, gender, relation to the HH head, ethnicity, marital status, occupation, constant income and source, and information on the members currently not living in the family. No changes were made to the section applied.

The **Education** section collects information on all members' education, starting from 7 years and older. Questions include educational attainment including type of school, number of

Table 4.1: Description of the Research Questionnaire Sections

No	Section	Information
I	Household Card (Profile information)	General household information
II	Education	Education status
III	Migration	Origination of the HH members
IV	Dwelling	Housing data
V	Availability of durables	Property information
VI	Agricultural activity	All production data
VII	Food expenditures	All expenses on food
VIII	Non-food expenditures and services	All non-food expenses
IX	Eating out, other sources of food	Food consumption out of the house
X	Loans and savings	Debts and savings data
XI	Non-agricultural activity and remittances	Non-farm employment and transfers abroad

Source: Author's research questionnaire

years of study, reasons for non-education for school-age members, level of degree and area of specialization (for adult members). Information on pre-schoolchild care and some more detailed information on additional training for school-age children was deleted. The reason was that kindergartens and the possibility for additional training are rare in most of the villages in the area.

The **Migration** block collects information about the nationality, place of birth, reason for migration to the current place of residence, whether members live anywhere else for more than 3 months, and current registration. There are no changes here too.

The **Dwelling** section collects information on the type of dwelling the household occupies, the number of rooms, ownership, construction features, and access to services (electricity, gas, water, etc). It also collects information on some additional qualitative data on the quality of services. Some questions relating to urban dwelling were excluded.

The **Availability of Durables** section collects information on the durable goods of the household. It includes the list of electric appliances, furniture, carpets, etc. A separate list describes automobiles, the year of purchase, and cost. The list of goods was shortened. Some qualitative data as year of purchase and cost were excluded to shorten the time of filling in the questionnaire.

The **Agricultural Activity** section is one of the core parts of the questionnaire. It collects information about all aspects of the agricultural production and consists of 9 sub-sections. It includes questions on the type of land the household owns, works on, the number of hectares of each type of land, the selling and leasing value of the land, the main source of irrigation for each type etc. Additional indicators, which describe the land's resources in more detail (quality of land through self-assessment, distance to the plot of land, water supply and costs) were introduced separately. The types of crops grown for each culture, the amount kept as seed, the amount sold, lost due to various reasons, consumed by the household, the amount put in storage etc.; the amount and cost of mineral fertilizers used, spending on various kinds of paid labor, separate for mechanized works and for manual operations, fuels, land taxes, livestock, other taxes etc. This section also collects information on food products as crops grown and produced by the household. A separate sub-section describes information on livestock, poultry, bees or other animals, the household's own products obtained from animals raised by the household, veterinary services, livestock expenditures such as feed, hired labor for herding, working animals and other farming equipment. The land quality sub-section was included here. The list of work depending on the regional dominating culture was also subject to change and a separate list for mechanized works and manual operations was added.

The **Food Expenditures** part collects detailed information on the type, amount and value of food items purchased for consumption during the past 12 months; the market channel type; the number of members residing constantly in the household's dwelling; purchase of food on credit. The list of goods was shortened by excluding unnecessary items or through grouping of similar products. Also excluded was information on the food purchased over the past two weeks, because this column in the LSMS questionnaire was filled in on the base of a two-week gap between the visits of the interviewer and could be skipped in our case.

The **Non-Food Expenditures and Services** section collects information on the various details of expenditures. It includes the various expenditures by items grouped by destination - non-food goods and services, costs for heating, utility services, health care expenditures, transportation costs, educational expenditures, and other expenditures including separate support for children, friends and relatives. Questions were rephrased for periodicity for the last quarter, in some cases for the last year.

Eating Out, and Other Sources of Food is a relatively small section collecting information about meals or snacks purchased and eaten outside of home or drinks taken outside of home by members of the household; it also asks for additional food received from friends/relatives and/or humanitarian organizations. Questions on additional food sources were included

The **Loans and Savings** block collects data on any loans and borrowing made by the family members within the last 12 months, values of loans and borrowing, payments, regularity, types of bank, and the rest of the loan; it also enquires about the household savings. Additional questions about the aim of the credit (land purchase, rural or commercial activity, purchase of animals etc.) were included.

The **Non-Agricultural Activity and Remittance** part collects information on the non-farm activities of the household members. Two different types of activity are presented here - hired worker or self-employed entrepreneurial activity. The questions include: type of employment, expenses and average income over the last 12 months; this section also includes a small block on household members working abroad and remittances received from them by the HH during the last period. This section was shortened to a one page format.

Except for the described changes, the four following sections were excluded from the questionnaire:

- health issues,

- family planning and female health,

- anthropometric data,

- employment and income.

The first three excluded sections were outside the scope of the research subject. However, information on health care expenses were incorporated in the section dedicated to non-food expenses and services. The employment and incomes section was deleted, because most of the activity of rural inhabitants is now concentrated in the rural sector, and the rest of the income activity is incorporated in the household card and in the last section, dedicated to non-agricultural activities and remittances.

4.2.2   Survey Sample Design

Main idea for the sample selection procedure was to represent all rural households in the selected target research area - Talas Oblast. Judging from the 2009 census, there were about 44,213 households, containing 225,757 individuals living in the Talas Oblast. Rural population consists of 35,358 households, including 191,102 individuals. Our target household sample size planed of 300 households. To allow fairly low non-response rate around ten percent, based on the experience of the previous pilot study in Fall 2010, it was drawn a sample of 330 households. It was close to final result of 297 collected questionnaires at the end.

Table 4.2: Rural Population of Talas Oblast and Selected PPs

	Population			Selected Population Points			
	PPs	person	%	PPs	persons	HH	%
Rural population	90	191008	100	30	72301	17116	38
villages with:							
< 5000 persons	8	52729	28	4	25928	4824	49
3001 - 5000 persons	13	47585	25	4	16456	1219	35
2001 - 3000 persons	14	33492	18	4	9829	3119	29
1001 - 2000 persons	30	43985	23	10	14862	5206	34
> 1000 persons	25	13217	7	8	5226	2747	40

Source: NSC Census, 2009

Table 4.3: Distribution of the Selected PP Across the Rayon Level

Rayons	Population			Distribution by size				
	PPs	covered	%	>5001	3001-5000	2001-3000	1001-2000	<1000
Talas	27	8	29	1	1	1	3	2
Kara-Buura	22	8	35	1	1	1	3	2
Manas	22	7	55	1	1	1	2	2
Bakai-Ata	19	7	38	1	1	1	2	2
Total	90	30	38	4	4	4	10	8

Source: NSC Census, 2009

Because of the regional character of the study it was possible to skip some of the procedures implemented at the national level (e.g., selecting primary sampling and secondary sampling units). In the region, the whole rural population lives in 90 villages located in four adminis- trative divisions - rayons. Each rayon also consist of micro-divisional entities - Aiyl Okmotu (see 3.1.3 on page 32). There are 36 Aiyl Okmotu in the four rayons. Due to the purely administrative character of these local entities it was decided to ignore them in order to simplify the random selection of villages. To save time and resources, it was reasonable to focus on the target group of the population. Table 4.2 presented the features of Talas Oblast's population and the selected target population points (PP).

90 villages were counted in the region, and we wanted to cover every third village. It was decided that 30 villages presented a reasonable task, and that it was possible to cover them

in the planned period. The selection of the villages was conducted in two steps. In the first step, all the villages were ranged by size into five categories (see the Table 4.2 on page 59). Then it was decided to select a certain number of villages from each category to cover an approximately equal share of the population in each category.

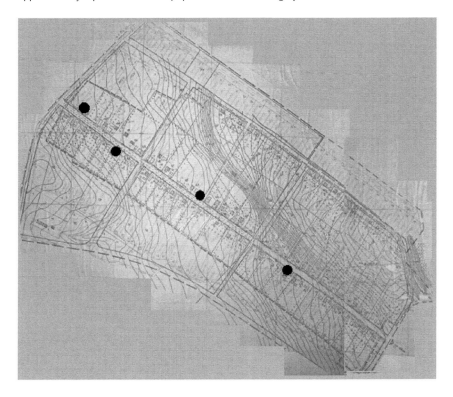

Figure 4.7: Example of the Geographical Sampling in the Sample Village
Source: Modified map image from the Talas Oblast State Land and Immovable Registration Service

The next step of sample selection (Table 4.3 on page 59) concerned the distribution of the selected population points (PP) between four rayons (districts) in the area. The selection was based on an equal distribution of the amount of villages between the rayons and according to the gradation of the villages by size. A certain difference between rayons is based on the different population density only

In the selection procedure at the rayon level, extremely big (more than 10 thousand people) or small villages (less then 400 people) and also several villages located in extremely complicated places (when a one way trip takes more then 3 hours) were excluded. Also, preference was given to a higher geographical distribution among objects within the rayon area, which means that villages of different size, which were located too close to each other, were excluded. In Talas Rayon, three villages that were covered in the pilot test survey were also excluded.

Distribution at the village level was based on an approximately equal randomization of around 2.5% of the households from the total of 13,467 households in the selected 30 villages. The expected sample was supposed to lead to a target sample of 300 households, covering around 2.2% of the households in the area. The covered population's share, however, differs in small and big villages. It varies from 3.5 to 5.3% in the small villages to approximately 2% in the big and medium sized villages (see Table 4.4 on page 63).

To strictly equalize the exact share of the HH in small and big villages, it was necessary to decrease the number of households in small villages and to increase it for the big ones. This decision, however, was not rational from the logistic point of view and in addition may have led to an excessive prevalence of the share of HH from big villages in the database. Thus for small villages the sample became approximately more representative then for bigger villages.

Sample selection at the village level was based on the geographical randomization of the households within the population point. The randomization was applied to the visual distribution of the selected HH by using village maps. Maps for all selected villages were obtained at the initial stage of the field survey. Sources of the maps were the rayon divisions according to the State Registration Service (Gosregistr).[I] The agency is responsible for technical registration of the land's resources and immovable property in the country. The permission to receive those maps was granted thanks to a letter asking for support of the research project, sent by the Government of the Kyrgyz Republic to the State Administration of Talas Oblast.[II] Maps for the village level only exist in paper form and typically consist of several separate sheets. Maps for one village may consist of 2 to 19 sheets of paper in the size of format A1. The maps were photographed with a digital camera and then consolidated on simple software (Paint) in one picture. Then the maps were used as schematic sample for the selection of the households situated within the territory of the village. Distribution of the sample within the village is based on the preliminary fixed intersection points (approximately one of forty houses) to randomly cover the area of the selected population points. For logical reasons it was decided exclude areas where no households were to be expected, such as commercial areas, local government, schools, parks, and other public and technical buildings.

A typical example of the method is presented in Fig. 4.7 on page 60. This small village with around 600 residents represents a typical small populated area with around 120 houses and a target of 4 covered households. All the houses are located in one street and the households

---

[I]More information on the agency is available at http://www.gosreg.kg
[II]Letter is attached in Appendix B.

selected are on the two opposite sides of that street. It was suggested to select the fifth house from the beginning of the street. The interviewers, however, could choose the house within the preliminary intersection point on the map (e.g., they might choose the side of the street and move in forward or reverse directions in case of refusal or absence of the house's residents). Maps for all the villages covered are available for study in Appendix C.

Interviewers were given no specific discretion in the selection of households; substitution or replacement was permitted in certain specified cases - absence of people in the household, destruction of the house, direct refusal of the household head to be interviewed. In these cases the interviewer could change the selected house within the range of the next five houses in both directions on both sides of the street (if available) around the originally selected house.

Thus, the methodology concentrated on keeping the distribution of the selected villages across all rayons of the region random in proportion to the population's concentration; the next stage was similarly based on the random distribution of the selected houses across the territory of the village.

4.3 Survey Organization

4.3.1 Survey Management and Preparation

The survey organization was conducted in two stages - the pilot (test) study and the full field study. The initial preparation for the field study started in summer 2010. The main task of the pilot visit was the preparation of the survey's main components - the questionnaire, and the sample method test. Additionally, it supported the preparation with regard to the local conditions, including staff selection and training, organization of logistics and housing, and arrangement of the cooperation with the local authorities, including the search for a possibility of obtaining maps.

Three villages were covered in a pilot survey in Talas Rayon - Kok-Oi, Kum Aryk and Kok-Kashat. 24 questionnaires were filled (12 in Kok-Oi and per 6 each in the other villages). During this process, the questionnaire were shortened and restructured several times. Initially the survey was implemented individually. Later, two persons from the staff of the local university and non-governmental organization (NGO) were involved additionally.

The training process became part of the work; unfortunately both options showed certain problems in the training process - understanding the questions, mistakes in the interview process and in filling in the questionnaire, and contact with the respondents. An alternative

Table 4.4: Distribution of the Target Sample Across Selected Population Points

Village	Population		Target	%	Expected	%
	People	HH		HH	HH	
Bakai-Ata Rayon						
Bakai-Ata	6834	1355	33	2.4%	30	2.2%
Ak-Dobe	4298	865	20	2.3%	18	2.1%
Min- Bulak	2729	476	11	2.3%	10	2.1%
Madaniat	1394	269	7	2.6%	6	2.2%
Kyzyl-Sai	1324	266	7	2.6%	6	2.3%
Namatbek	803	159	4	2.5%	4	2.5%
Urmaral	460	89	4	4.5%	4	4.5%
Kara-Buura Rayon						
Amanbaevo	6203	1123	25	2.2%	24	2.1%
Chimgent	4990	901	20	2.2%	15	1.7%
Bakyan	2585	443	11	2.5%	10	2.3%
Jon-Dobe	1860	349	9	2.6%	8	2.3%
Suulu-Maimak	1716	311	8	2.6%	7	2.3%
Uch-Bulak	1479	308	7	2.3%	6	1.9%
Kok-Dobe	647	117	4	3.4%	4	3.4%
Tamchi-Bulak	441	75	4	5.3%	4	5.3%
Manas Rayon						
Pokrovka	7419	1401	33	2.4%	30	2.1%
Talas	3319	610	15	2.5%	13	2.1%
Kyzyl-Jyldyz	2421	442	11	2.5%	10	2.3%
Balasary	1865	352	9	2.6%	8	2.3%
Jaylgan	1337	252	6	2.4%	6	2.4%
Jiide	841	153	4	2.6%	4	2.6%
Kengesh	762	139	4	2.9%	4	2.9%
Talas Rayon						
Kopuro-Bazar	5472	945	22	2.3%	20	2.1%
Aral	3849	710	17	2.4%	15	2.1%
Sasyk-Bulak	2094	413	10	2.4%	10	2.4%
Orto-Aryk	1453	243	6	2.5%	7	2.9%
Kozuchak	1335	246	6	2.4%	5	2.0%
Arashan	1099	203	5	2.5%	4	2.0%
Kenesh	695	137	4	2.9%	4	2.9%
Chong-Tokoi	577	114	4	3.5%	4	3.5%
	72301	13467	330	2.5%	300	2.2%

Source: NSC Census, 2009; NSC (2010a); Survey Data

source of field interviewers might have been the local Statistical Office, but a cooperation was not possible due to the higher price for its services.

The full field study started in May and covered the entire region. Field forms translations of the questionnaire were completed and printed in Kyrgyz. Due to the previous experience from the pilot phase, the field interviewers' supervisor was assigned to Bishkek at the initial stage on a full-time basis for the five-week period of data collection. The other three interviewers were found and assigned locally to Talas Oblast. Operation control and management was arranged through personal involvement in the process of selecting the staff, organizing the preparation of materials, and in procurement and logistics, as well as instruction and learning-by-doing.

An important element of the survey was the arrangement of access to the village maps drawn up by the local State Registration Service Offices. During the initial stage, maps had already been prepared for Talas Rayon as a result of the pilot project. Thus, it was necessary to get the maps of the three other rayons. It was decided to arrange preliminary available maps from the open source "Google Available Maps" (see Fig. 4.8 on page 65). The accuracy of these maps was sufficient for selecting preliminary intersection points for the selected target households. In the end, all maps with the households that had actually been reached were checked additionally. This gave us the possibility to be flexible in the village selection; otherwise it would have been necessary to take photos of the maps for all of the ninety villages. Maps from other rayons were collected at the end of the survey and compared to eliminate possible mistakes and to mark the households selected by the interviewers.

Transport became an important component of the project. In the survey process, selected villages were changed - in some cases because of problems with the roads and the great distance that had to be travelled to reach the destination. Because of the strong support from the contacts in the Central Government Office additional transport support from the local state administration was arranged. Transport with a driver was provided for the trips to the remote villages, except for the cost of fuel. This decreased the operational costs for transport, and was helpful too from the safety point of view. In one remote village there was a problem with aggressive behavior of one of the local residents. This incident did not allow us to finish the planned sample in the village. However, thanks to our local driver provided by the state organs a bigger problem was prevented. It was only this one incident which compromised the safety of the team.

In order to optimize logistics and the organization of work, it was decided to set four questionnaires as the smallest possible number of collected questionnaires in a village. A bigger number of the households, for example in a very small village, might have led to an oversampling of the small village population, while a smaller amount of covered households would have led to an increase of the operational costs per questionnaire for a small village.

Figure 4.8: Example of the Preliminary Sampling in the Village of Bakyan, Kara-Buura Rayon
Source: Google Maps

4.3.2 Staff Training and Quality Control

A special operation manual for interviewers was not developed. Instead, the interviewers underwent an intense training procedure. Each of the interviewers worked together with a supervisor for three days. The first day was dedicated to the study of the questionnaire and to the technique of filling it in. On the second day, the interviewer started to work under supervision and after each interview he / she was corrected with regard to possible mistakes. As of the third day the interviewer started to work independently, but after each interview the questionnaire was checked for mistakes.

The following aspects of interviewing were taught in detail:

- the use of maps of the village (Gosregistr and Google maps) to select the target house,

- entering and starting the interview, answering the survey goals and explaining the task,

- filling in the questionnaire content correctly, conversation techniques,

- finishing the interview and leaving the household

Table 4.5: Actual Survey Schedule, April-July, 2011

	Months	April	May			June			July
	Days	21-30	1-10	11-20	21-31	1-10	11-20	21-30	1-10
I	Preparation								
1	Questionnaire	x	x						
2	Staff selection	x	x		x				
3	Organization								
	of field work	x	x	x	x	x	x	x	
II	Field Survey								
	Implementation								
1	Start of the process		x						
2	Training of Staff		x	x	x				
3	Interviewing		2	50	86	78	32	49	
	Households (297)								
4	Quality control		x		x		x		x
5	Field part finalizing							x	
III	Finalization								
1	Reporting								x
2	Data entry start								x

The interviewer was able to ask by mobile for interpretations of some of the questions which were part of the interview process; it took around a week to be fully involved in the process. Not all interviewers were equally successful in achieving a sufficient level of confidence. There was a certain problem with discipline and accuracy. Some interviewers left the process early, and then the process of training a new interviewer was repeated again. Four persons in total, except for the supervisor (researcher), were involved in the survey interviewing process.

Quality control procedures were arranged through personal training, daily operational presence in the field, and checking of the filled-in questionnaires to correct mistakes. Only trained interviewers worked independently in the small and medium-sized villages. Altogether nine villages were covered independently by the trained interviewers. In some cases (three villages - two interviewers) verification via field visits to selected households was implemented before payments were made. Those three villages were Jiide (Manas Rayon), Chong-Tokoi (Talas Rayon), Namatbek (Bakai-Ata Rayon). During the final stage, control visits were made by the supervisor to check whether the HH in those villages had been visited.

Table 4.6: Distribution of the Completed Target Sample Across Selected PPs

Villages	Questionnaires	Villages	Questionnaires
Bakai-Ata Rayon	**78**	**Manas Rayon**	**73**
Bakai-Ata	30	Pokrovka	28
Ak-Dobe	18	Talas	14
Min- Bulak	10	Kyzyl-Jyldyz	10
Madaniat	7	Balasary	7
Kyzyl-Sai	6	Jaylgan	6
Namatbek	3	Jiide	4
Urmaral	4	Kengesh	4
Kara-Buura Rayon	**76**	**Talas Rayon**	**70**
Amanbaevo	23	Kopuro-Bazar	20
Chimgent	15	Aral	15
Bakyan	9	Sasyk-Bulak	10
Jon-Dobe	8	Orto-Aryk	7
Suulu-Maimak	7	Kozuchak	6
Uch-Bulak	6	Arashan	4
Kok-Dobe	4	Kenesh	4
Tamchi-Bulak	4	Chong-Tokoi	4
Total		297	

Source: Survey Data

4.3.3 Project Schedule Implementation and Final Data Collected

The actual schedule of survey implementation can be found in Table 4.5 on page 66.

The final data collected for the survey are as follows:

Total sample households selected	330
Minus HH found to be vacant	3
Minus HH found to be demolished	16
Minus HH not eligible for other reasons	5
Total sample households eligible for interview	306
Minus HH that refused to be interviewed	7
Minus HH that could not be contacted	1
Minus HH that did not respond for other reasons	1
Total households which completed an interview (97%)	297

The distribution of completed households across the rayons and villages can be seen in Table 4.6 on page 67. Geographical randomization of the covered villages is presented on the map in Fig. 4.9 on page 69.

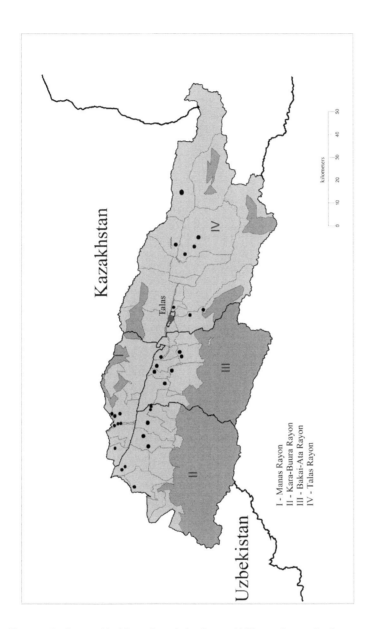

Figure 4.9: Geographical Location of the Covered Villages Across Region

Source: Author's own presentation from map of Talas Oblast Census 2009, NSC

69

5. Descriptive Statistics

This chapter presents descriptive statistics of the field study sample. It explains the main sample characteristics and specific features of the agricultural production on the household level and other activities and income sources of the rural inhabitants.

5.1 Household Sample Profile

The sample consists of 297 households, located in 30 sampled villages distributed across the region. 1616 persons are listed as members of the households. Of those, 1600 persons actually presently live in the residential places, while 16 persons from 14 households were temporarily absent during the survey period, but formally belonged to the visited households. The average size of the household in a sample is 5.44 persons, similar to the Oblast level (NSC, 2010a, p158).

Table 5.1: Main Characteristics of the Sample in Comparison With Talas' Rural Population

		Sample Data			Oblast Total		
		Total	Male	Female	Total	Male	Female
1	Population age groups:						
a	children (0-14), %	35.4	35.7	35.0	34.3	34.8	33.8
b	working age, % (men 16-59, women 16-54)	55.4	58.8	51.7	55.6	57.6	53.5
c	non-working age, %	9.2	5.5	13.2	7.7	5.2	10.3
2	Average age, years	27.3	26.4	28.3	26.5	25.8	27.2

Source: Survey Data, National Census Data NSC (2010a)

In the sample, the share of the male gender slightly prevails (51.8% of men against 48.2% women) In the total rural population of Talas Oblast the prevalence of men is less obvious (50.4% male residents in comparison with 49.6% female population) (NSC, 2010a, p3). The difference can be seen in more detail in Table 5.1 on page 71, where the main age groups of the population and of the sample are compared in total and by gender structure[1]. Since demographic trends did not change markedly in that short a time, other explanations, for

---

[1] As of 2011 the age of retirement is 63 years for men and 58 years for women

example, the change of the gender pattern in the course of time, are not realistic. This difference may be explained primarily by the random sampling procedure of the survey. The general pattern of the sample coincides with the regional rural population age pattern, but certain differences exist. The sampling error effect may also be seen in the more detailed age and gender pattern of the covered sample in comparison with the total population structure presented in Figure 5.1. Despite visible deviations of the distribution of the smaller age groups, the differences reach at most 1.9% of the group share in the structure of the gender strata, but in general do not exceed 1%. Nevertheless, differences do exist and one certainly detects a sample measurement error.

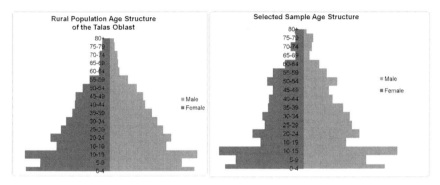

Figure 5.1: Age Structure of the Sample and Total Population of Talas Oblast
Source: Survey Data, National Census Data NSC (2010a)

Another general trend we could positively identify was a significant share of the population who are of working age. Another related trend is the increasing birth rate in Kyrgyzstan, observed since 2000. It stems from the high birth rate of the population in the 80s and beginning 90s. The increase of the group of women who reach a favorable reproductive age (of 20 to 29 years) directly led to the growth of the population over the last decade (NSC, 2011a, p6-7). The decrease of the birth rate in the 90s led to the following shift of the group of young people from 16 to 19 years of age. These deviations are typical for the whole country, but were specifically evident in Talas and Naryn Oblasts.

The national structure of the sample is representative (see Figure 5.2), which is mainly due to the region's high level of ethnic homogeneity. The structures of the sample and the level of the whole region are almost identical. Difference in small ethnic group shares are insignificant.

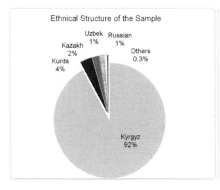
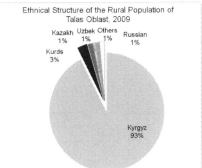

Figure 5.2: Ethnical Structure of the Sample and Total Population of Talas Oblast
Source: Survey Data, National Census Data NSC (2010a)

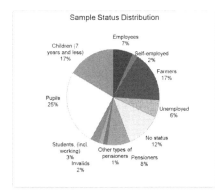
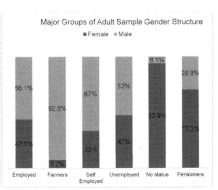

Figure 5.3: Structure of the Sample by Status Groups and Gender Structure
Source: Survey Data, National Census Data NSC (2010a)

Household heads are primarily male representatives of the family (86%) with an average age of 49 years. Of female heads of households, which make up the remaining 14%, the majority (75%) are widows, with an average age of 56.7 years. The fact that women can very often only became head of a HH after the death of a husband demonstrates the patriarchal character of the family construction. This conclusion is also supported by the analysis of the household members status in Figure 5.3 (see p. 73).

Data from the right slide of Figure 5.3 demonstrates a difference in the status of the women and men in the sample. Men prevail in the groups 'Farmers' and 'Self-employed', while women lead in the group 'No status' and 'Pensioners'. The leading position of women in the group of retired persons can be explained by their higher longevity, but the prevalence of the women in the group of persons without a clear status needs to be studied separately. The groups denoted 'Unemployed' and 'Employee' are not characterized by a strong prevalence of one gender. 26% of the sample are occupied by the active population, primarily males, while 18% of the people in the sample specify themselves as 'Unemployed' and 'No status' persons, with a mainly female contribution. 11% of the sample consist of pensioners and invalids. A significant share of the population (45%) is occupied by children of up to 7 years, pupils and students.

5.2 Agricultural Production Activity

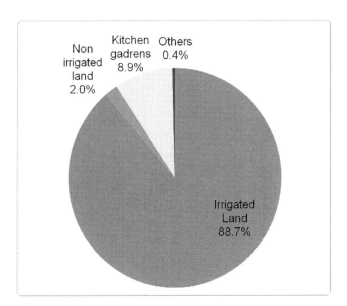

Figure 5.4: Land Resources Structure
Source: Survey Data

At the household level, all agricultural activity in Tales Oblast can be divided into two basic activities - crop production and livestock breeding. Most often, households combine both of these production activities, but usually with a special concentration on a certain type. However, in some cases the activity is limited due to an absence or lack of production resources, another type of activity or the inability of the household to afford production activity.

Land

Five main land types exist in the land ownership of the area's rural households - irrigated land, non-irrigated land, hayfields, gardens and kitchen garden plots. The structure of land resources is given in Figure 5.4. Irrigated land plots are the most important and biggest land resource in the area. 88.8% of the households own irrigated land, with an average land plot of 2.07 ha per HH. The average size of land plots is biggest in Kara-Buura Rayon (2.57 ha), while smallest are found in Manas Rayon (1.66 ha). The share of households without land plots is biggest in Manas Rayon (32%), in other rayons it varies from 4 to 8%. This is mainly linked with the density of the population in the respective area. The second land resource is the land plot near the house - the kitchen garden. 99.5% of HH have kitchen gardens, the average size is 0.184 ha per HH. Only in Manas Rayon, there were four households without a kitchen garden. The kitchen garden typically plays the role of a basis for the household's own food consumption, usually fruits and vegetables, while other types of land, intended for the growth of market products, are located at some distance away from the dwellings.

Not all households use the land resources of their own land for production. Of 540 hectares of available irrigated land 8% are rented out by 25 households. 36% of the rented land is located in Manas Rayon. Besides renting out in two cases (3.6 ha) land plots belonging to households were transferred to the children for free exploitation.

In contrast, 36 households rent additional tenure land plots, with a total area of 64 hectares. 6 of those households do not have their own land plots and thus overpass this problem. Additionally rented land types are: irrigated land (81%), non-irrigated land (16%) and hayfields (3%).

Crop Production

The agricultural production on the main land plots is primarily dedicated to the production of the market products or forage crops, while products produced in the kitchen garden are mainly consumed by the household. 570.4 hectares of the main land plots, including irrigated land, non-irrigated land and other types of land resources, are used for crop production. Four crops cover 96% of the used land plots - haricot beans, hay (lucerne), wheat and potatoes (see Table 5.2).

Haricot beans are extremely dominant in Bakai-Ata and Kara-Buura Rayons (87% and 94% of land respectively). Hay play the role of main forage crop in Talas and Manas Rayon, but is also used in other rayons as a crop rotation culture in order to recover land quality. Wheat and barley are used primarily for forage grain production, but in some cases wheat is used for food purposes. Potatoes are produced for commercial purposes in Talas Rayon,

Table 5.2: Main Crop Distribution in the Sample in Total and by Rayons (ha)

	Crops	Total	Rayons			
			Talas	Bakai-Ata	Kara-Buura	Manas
1	Haricot beans	319.3	24.8	129.2	139.9	25.3
2	Wheat	49.8	18.3	-	18.9	12.6
3	Hay (lucerne)	135.1	65.0	12.6	27.2	30.4
4	Potatoes	40.5	29.9	6.0	3.7	0.9
5	Barley	13.2	11.2	-	2.0	-
6	Other crops	12.6	0.1	1.6	3.4	7.5
	Total land	570.4	149.4	149.4	195.0	76.7

Source: Survey Data

in other rayons its share is minor. Other crops, playing minor roles in crop production, are corn, sugar beet, apples, sunflower seeds, onions and sweet pepper.

Land distribution by crops as seen in the sample is different from the overall picture described in the previous chapter (see Fig. 4.3 on page 50). Again, the reasons for that may lie in the random selection of the sample, but an alternative explanation could be that rural production on the household level covers only 75% of the land resources. The remaining quarter of land belongs to the local authorities and is basically tenured to big local farmers.

Table 5.3: Main Crop Productivity in the Sample in Total and by Rayons (ton per ha)

No	Crops	Total	Rayons			
			Talas	Bakai-Ata	Kara-Buura	Manas
1	Haricot beans	1.5	1.8	1.6	1.3	1.3
2	Wheat	2.0	2.0	-	1.8	2.3
3	Hay (lucerne)	1.9	1.8	2.8	1.8	1.6
4	Potatoes	9.3	9.4	7.5	11.5	8.9
5	Barley	1.9	2.1	-	0.8	-

Source: Survey Data

The overall land productivity declared by the surveyed households is lower than that shown in official statistics (compare data in Table 5.3 and Fig. 4.5 on page 52). While the difference is not that big for wheat and haricot beans, productivity in the case of potatoes is almost two times lower than shown in official statistics. The survey data appear to be more realistic, however.

To evaluate the total value of the crop production, the volumes of the traded product, non-traded product (intended for seed, consumption and stock), and barter exchanges were defined. Another issue is the reported losses for different reasons of the potential harvest. Five households informed us about the losses - in four cases a loss of about 2.6 tons of haricot beans was noticed and once it was 1.5 tons of wheat. We suppose that actual losses due to numerous reasons were substantially higher, but in the specified cases it was substantially visible for the farmers. Finally, losses of the product were not included in the final evaluation of the crop.

The volume of forage crops dedicated to livestock feeding was also not included in the volume of crop production, except for the cases of sale on the market. The rationale for this decision was their presence in the final evaluation of the livestock production in the form of indirect costs. Thus, one way or the other, the cost of forage is reflected in the overall agricultural production of the household.

In each household, the price for traded products was defined as the sum of money received for the volume of sold product. However, for the assessment of the value of non-tradable crops, the households' own consumption and barter exchanges, prices were aggregated in order to include them in the final product. Applied were the average prices (or arithmetic mean) for each product derived from the total sample, and data was taken from the sales price of the product and calculated as follows:

$$\mathbf{P}_i = \frac{\sum_{j=1}^{n} \mathbf{P}_{ij}}{n}, \tag{5.1.}$$

where P_{ij} is the price of the jth - sale of the ith crop product, and n is the amount of sales of the i-th crop product. We do not use the weighted average price, because in this case specialized farm-households selling the majority of their product on the market will have a bigger effect on the price than small subsistence-level households selling minor volumes of the same product.

The majority of the crop products by volume were sold on the market (72%). The total value of the sold products was 18 million Som [II]. 70% of this sum were provided by haricot beans. The second place, with 22%, is held by potatoes. Other products played a minor role. Non-traded crop products were evaluated with a sum of 1.4 million som. There were only four cases of barter exchanges, with a total value of 50 thousand som.

Kitchen gardens are the basis for a wider assortment of crops. 29 different crops were grown by the households, mainly fruits and vegetables. In the assessment of this micro-scale production, certain measurement problems arose. In general, people cannot define the exact

---

[II]The average exchange rate of Kyrgyz Som in 2010 was 45.96 som per 1 US dollar (NSC, 2012b, p1)

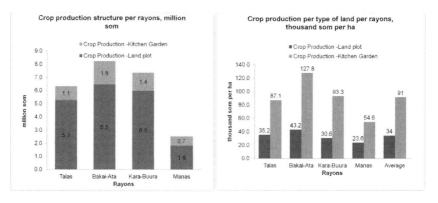

Figure 5.5: Value of the Crop Production per Type of Land and per Unit of Land
Source: Survey Data

area under a certain culture, therefore area measurement was skipped. But people were able to recognize the product range, the amount of a product, and how much of the product was consumed and/or sold on the market. The main products were potatoes, apples, cabbages, carrots, onions and tomatoes (83% of the total home production value).

Products from the kitchen gardens were not always traded. 11 products of 29 were used for personal consumption only and local prices were not available. Prices for 18 products with the available sale information were calculated with the same formula (formula 5.1. on page 77). Alternative prices were suggested for the rest of the products. Prices for two of the products (hay and sunflower seed) were taken from the average calculated price of the main land plot. Prices for the next three products (melons, water melons and walnuts) were taken from the average price of the purchased products consumption data set (formula 6.1.; further explanations are given in the poverty measurement section on page 93). Prices of the remaining products (cabbage, pumpkin, eggplant, raspberry, radish and grape) were taken from the average market price data of the NSC for 2010 (NSC, 2012a, p29-36). Prices for 2010 were taken because the yield assessment defined in the questionnaire was produced in 2010, and at the time the survey was conducted a new harvest, even from the kitchen gardens, was in the process of ripening. The value of these 6 products was equal to 3.3% of the total production of the kitchen gardens. Similar to the main land plot, the value of produced maize and hay dedicated to the feeding of animals was excluded from the kitchen gardens' final production. Finally, the total value of the crop products produced in the kitchen gardens amounted to 4.6 million som. Thus, the total value of the crop product of the main land plots and kitchen gardens was 24.2 million som, with the average crop production in the area of 81.4 thousand som per household.

Distribution of the crop production across the territory demonstrates that the main crop centers are Bakai-Ata and Kara-Buura Rayons (see left slide in Fig. 5.5). The lowest production capacities were shown in Manas Rayon. At the same time, the agricultural production pro-ductivity measured per unit of land (see right slide of the Fig. 5.5) shows that even the productivity in Manas Rayon was lower than average, the difference in comparison with the other parts of the region is also provided by the lack of land resources. The productivity per unit of land also demonstrates that kitchen garden produce yielded three times more product on average than the main land plot located in the fields. The reasons for this do not simply lie in a higher productivity due to a better gardening performance (e.g., picking weeds or better watering), but also in the higher variety of more expensive products, mainly fruits.

Livestock Production

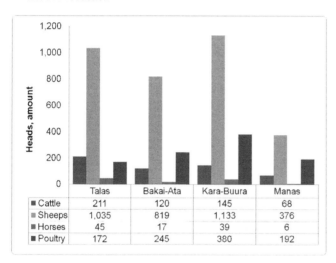

	Talas	Bakai-Ata	Kara-Buura	Manas
■ Cattle	211	120	145	68
▨ Sheeps	1,035	819	1,133	376
▥ Horses	45	17	39	6
■ Poultry	172	245	380	192

Figure 5.6: Livestock Distribution by Types of Animals per Rayons
Source: Survey Data

Livestock production in the area is based on four major types of livestock - cattle, sheep, horse and poultry. Other animals, rarely observed, are donkeys, goats, pigs and rabbits. 29% of the observed households could not afford livestock yarding. 544 head of all types of cattle, including 324 cows, were reported by 59% of the respondents. 3.3 thousand sheep were kept by less than half of the households. Poultry, less than 1000 heads, were found in 32% of the households. The maintaining of horses was only possible for 17% of the observed households.

Livestock is distributed non-uniformly throughout the territory of the oblast (see Fig. 5.6 on page 79). Manas Rayon showed the lowest level of livestock with regard to the main types of animals. Talas and Kara-Buura Rayon competed for domination in the field of animal breeding thanks to the availability of pastures. But in both of the rayons the leading position was achieved by extreme outliers. In Talas Rayon one household reported about 86 heads of cattle, while in Kara-Buura Rayon again one household counted 160 heads of sheep. However, even without those outliers, both rayons showed a higher level of livestock concentration. With regard to sheep livestock breeding there was a certain concentration in the level of the livestock when an economy of scale became visible. 21 households with sheep herds starting from 40 heads owned 40% of the total amount of sheep. In the case of cattle or horse breeding, those households were recognized as outliers.

Three types of income/product were extracted by the households through livestock breeding:

- sale of animals,

- livestock capital value increase,

- livestock products production:
   [a] for the HH's own consumption,
   [b] for sale.

Sales of live animals are the main source of the cash income from livestock production. Cattle, sheep and horses provide 99% of all sales. With a 46.7% share, cattle sale demonstrates a significant difference in sales of its separate sub-groups (cows, calf and bulls). While the amount of animals sold over the last 12 months compared to the actual amount of animals sold at the time of the survey reached 15% for cows sold annually, this share increased to up 59% for calves and more than 100% for bulls. The situation is different for horses, where mainly adult horses are sold. Looking at sheep, a distribution by subgroups is not possible, because in the interviews most of the residents made no distinction with regard to subgroups, but usually gave the total amount of sheep instead. Sheep sale takes second place with 43.8%. Annually, a quarter of the sheep is sold. Horses, with a 8.6% share, are in the third place.

As expected, the distribution of animal sales across the territory is unequal (see Fig. 5.7 on page 81). Because of cattle, Talas Rayon leads with 39% of all sales. Kara- Buura Rayon is in the second place with 32% of livestock traded, mainly sheep. In each case, the price of sale is provided individually on the household level. The final value of the sold animals reaches 7.2 million som.

The *livestock capital value increase* means a positive change of the amount of livestock, measured as the difference between the amount of livestock 12 month before and the current amount of animals at the time of the survey. The price for each type of livestock P is

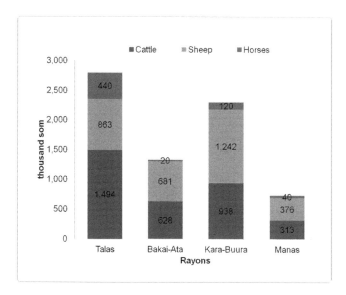

Figure 5.7: Livestock Sales Distribution by Types and Rayons
Source: Survey Data

calculated as the self-assessed price of livestock divided by the actual number of animals at the time of the interview period. The difference in the increase of animals is calculated by the following formula:

$$\mathbf{L_{cap_i}} = \frac{N_i - N_{i_{(12month)}} - N_{i_{purchase}}}{P_i}, \tag{5.2.}$$

where N_i is the number of ith animals during the survey period, $N_{i_{(12month)}}$ is the amount of the ith animal 12 months ago, $N_{i_{purchase}}$ is the amount of the purchased i^{th} animal for the last 12 months, and P_i is the price of the ith livestock. Thus, only the increase of the new issue and changing of the category of animals is counted here, e.g. from calf to cow / or bull.

The distribution of animal growth of the livestock across the territory is characterized by a changing trend in comparison with the sale of live animals (see Fig. 5.8 on page 82). Talas Rayon's leadership is based on the increasing sheep value, while in Bakai-Ata and Kara-Buura Rayon the share of cattle and sheep growth is almost equal. The share of Kara-Buura Rayon is smaller than in the previous category. That means a higher sales level in comparison with the livestock reproduction rate. Manas Rayon typically, is in the last place, but it is interesting to note that the increase of cattle capitalization is high enough if we take into account the low amount of cattle. The total value of the capital of animals increased

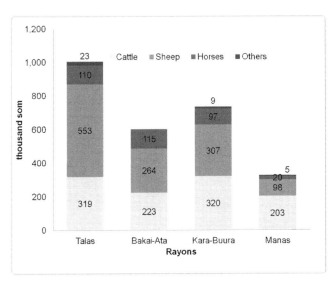

Figure 5.8: Livestock Capital Increase by Types of Animals and Rayons
Source: Survey Data

to 2.6 million som within the last 12 months.

Livestock products produced by households are subdivided by purpose, based on whether they are intended for the households' own consumption or whether they are intended for sale. The total range of livestock products produced by the households is limited to lamb, beef, eggs, chicken meat and milk. In single cases, the production of animal skins and pork meat can be found.

Sale is dedicated to the following products: beef, eggs, milk, pork meat and animal skin. The dominating position in this range is occupied by milk with 94% of all sales (see left slide in Fig. 5.9 on page 83). Other products play a minor role. It has to be pointed out that 76% of the produced milk is designated to be sold. Milk sales are mainly provided by Talas Rayon with 42% of sales (see right slide in Fig. 5.9 on page 83). The other rayons' shares are presented typically - Bakai-Ata and Kara-Buura Rayons show approximately equal shares (22% and 24%, respectively), while Manas Rayon again demonstrates the smallest share of 12%. The total sale of livestock products consists of 4.3 million som. The market character of milk production is evident. In every household, the price is individual, despite low variation, and an average price is not calculated.

Own-consumption livestock products include lamb, beef, eggs, chicken meat, milk and pork meat (one single case). It is interesting that lamb and chicken meat is only designated for the households' own consumption. Average prices were applied for self-consumed products. For

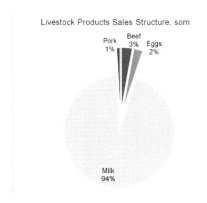
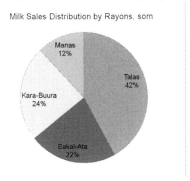

Figure 5.9: Structure of Livestock Product Sale and Milk Sales by Rayons
Source: Survey Data

products sold on the market formula 5.1. was used. Prices were available for beef, eggs, milk and pork. The market prices for lamb and chicken meat were unavailable at the local level. Therefore, their prices were taken from the average price of purchased products according to the consumption data set (formula 6.1. on page 93 was used to derive the average prices for lamb and chicken meat from the data set). In the literature it is recommended to use farm gate prices instead of market prices (Deaton and Zaidi, 2002, p18), but in this case any type of farm gate price was unavailable. The average price of NSC for the observed period was slightly higher than the average price for meat purchased by the rural households and we decided to use our own calculated prices.

Mainly consumed livestock products were lamb and milk with 59% and 29%, respectively (see left slide in Fig. 5.10). Eggs were in the third place with 8%, other products were not significant.

The highest lamb consumption was observed in Kara-Buura Rayon, while Manas Rayon demonstrated a level more than two times lower (see right slide in Fig. 5.10). Milk consumption was also significantly higher in Kara-Buura Rayon, while in the other three rayons the difference stayed within a smaller range. The value of the own consumed livestock products consisted of 4.5 million som.

The total livestock production volume was 18.6 million som, with an average production of 63 thousand som per household. Distribution of the livestock production across the territory showed a prevalence of Talas and Kara-Buura Rayon (see Fig. 5.11 on page 85). A significantly lower level of livestock production was found in Bakai-Ata Rayon, while livestock production capacities of Manas Rayon were more than two times lower in comparison with

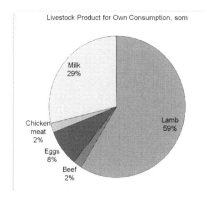
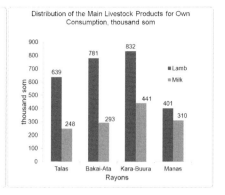

Figure 5.10: Structure of Consumed Livestock Products and Main Consumed Products Distribution by Rayons

Source: Survey Data

the leading rayons. Regional specialization had a visible impact on livestock production across the target territory.

Besides purely agricultural production it is possible to include revenues from the renting out of land resources in agricultural income. Renting out land plots brought 193.5 thousand som for 25 households, with an average rent of 4.3 thousand som per hectare. The highest rent was observed in Bakai-Ata Rayon (8.5 thousand som per ha) and the lowest surprisingly, in Manas Rayon (2.8 thousand som per ha), if we take into account the scarcity of land in this part of the region.

The *overall agricultural production* in the sample was assessed at a total of 43.3 million som. The crop production share consisted of 56.4% of income, while livestock production brought 43.2%, and renting of land 0.4%. More detailed information on the agricultural product structure across the territory is given in Table 5.4.

The regional distribution of agricultural products is seriously affected by the distribution of land resources. Manas Rayon demonstrates an extremely low level of production volumes in general (only 11% of agricultural product) and in production per household mainly due to the lack of production land resources and pastures. Bakai-Ata Rayon generally concentrates on crop cultures (the livestock sector shows 33%) and, predominantly because of haricot beans, has kept sufficient production capacities. Talas, Manas and Kara-Buura Rayons show a higher influence of the livestock sector (51%, 48% and 44%, respectively), but both sectors are represented. Haricot beans and cattle are the main pillars of production in the area, with 29% and 23% of the total product, respectively. Potatoes and sheep are the second

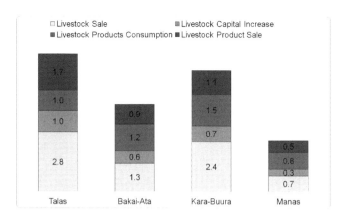

Legend:
□ Livestock Sale   ▨ Livestock Capital Increase
■ Livestock Products Consumption   ■ Livestock Product Sale

Chart bars:

Talas: 1.7, 1.0, 1.0, 2.8
Bakai-Ata: 0.9, 1.2, 0.6, 1.3
Kara-Buura: 1.1, 1.5, 0.7, 2.4
Manas: 0.5, 0.8, 0.3, 0.7

Figure 5.11: Livestock Production Structure by Rayons, million som
Source: Survey Data

Table 5.4: Agricultural Production in the Target Sample by Rayons

		Rayons			
		Talas	Bakai-Ata	Kara-Buura	Manas
1	Agricultural product total, million som	12.8	12.3	13.0	4.9
2	Agricultural product per household, thousand som	183	158	171	67
3	Structure of agricultural product				
a	Crop production - Main land	41%	53%	46%	37%
b	Crop production - Kitchen garden	8%	14%	10%	14%
c	Livestock and livestock products sale	35%	18%	27%	26%
d	Livestock capital increase	8%	5%	6%	7%
e	Livestock products consumption	8%	10%	12%	16%
f	Income from rent of the land	0.2%	0.6%	0.3%	0.9%

Source: Survey Data

most important production directions with a 16% share each. The highest per household production is observed in Talas Rayon, mainly due to a higher share of livestock. It is necessary to note that the highest level of livestock capital increase, which is neither the monetary income, nor a consumption product, has been detected in Talas Rayon. For those interested, additional information about the agricultural production features of the sample is presented in Appendix D.

5.3 Non-Agricultural Activity Statistics

Despite the priority of the agricultural production in the area, there are complementary sources of income and other types of activity. 82% of the households receive at least one of the following types of income:

- salary from permanent work,

- social transfers:

    [a] pensions,

    [b] social payments for children,

- income from the private sector:

    [a] hired workers,

    [b] income from entrepreneurial activity,

- remittances from transfers abroad.

Table 5.5: Salary from Permanent Work

		Total sample	Rayons			
			Talas	Bakai-Ata	Kara-Buura	Manas
1	Households	82	17	27	14	24
2	Persons	108	21	37	19	31
3	Total payments, million som	5.8	1.3	1.7	1.2	1.6
4	Average monthly salary, thousand som	4.5	5.3	3.8	5.2	4.3

Source: Survey Data

Salary from permanent work primarily stems from employment in the state and public institutions in the area, e.g., regional state services, hospitals, schools, post offices and other public services. However, this category also includes workers of big and medium businesses - energy and mining sectors, food processing, construction, financial institutions. More than a quarter of the sampled households (27%) reported that 12% (or 108 persons of 895) of the working age population were employed in constant work places (see Table 5.5 on page 86). Bakai-Ata and Manas Rayons demonstrated higher levels of permanent work places. It has to be said that both of these rayons, unlike two other rayons, are covered by villages - rayon centers, where local authorities offices are presumably concentrated. The size of the average monthly income from permanent work was equal to 4.5 thousand som per person, which might be assessed as the equivalent of the average crop production of 1.6 hectares of land.

The next most important source of rural cash revenues is the main social transfer of the rural population - pensions. More than 40% (or 130 of 297 sampled HH) of rural households indicated that they have this source of income (see Table 5.6). Besides all of the population of non-working age there exists special type of pensions, like disability pensions, pensions for special conditions, military service pensions, etc. The distribution of pension revenues across the territory is more consistent than the distribution of salaries from permanent work. Pensions in Talas Rayon are slightly higher than in the other parts of the oblast.

Table 5.6: Social Transfers I - Pensions Distribution

		Total sample	Rayons			
			Talas	Bakai-Ata	Kara-Buura	Manas
1	Households	130	27	37	36	30
2	Persons	173	37	51	48	37
3	Total payments, million som	5.5	1.3	1.6	1.4	1.1
4	Average monthly pension, thousand som	2.6	3.0	2.6	2.5	2.5

Source: Survey Data

In addition to pensions there are additional social payments dedicated to the support of children - scholarships and social payments for children from big families (see Table 5.7 on page 88). Scholarships for certain groups of students still exist despite their rareness. Only four persons were detected among all of the respondents, who received scholarships. There is no possibility to track a source of payment of these scholarships. The average size of these payments is modest and cannot sufficiently support a study process of the students.

Social payment for children from big families exists for families with three and more children, whose parents may prove absence of other sources of income. From the field interview experience we need add that financing of this category of social payment is quite strict and, in addition, social service representatives check the living conditions of the families, e.g., respondents stated that the presence of a TV or washing machine was the basis for refusal of such payments. Only thirteen households reported this type of social support (see

Table 5.7). The size of this type of compensation is really small and cannot be qualified as sufficient support for such families.

Table 5.7: Social Transfers II - Scholarships and Social Payments Distribution

		Total sample	Talas	Rayons Bakai-Ata	Kara-Buura	Manas
I	Scholarships					
1	Persons	4	2	-	2	-
2	Total payments, thousand som	43.0	20.4	-	22.6	-
3	Average monthly income, som	895	850	-	940	-
II	Social payments for children from large families					
1	Households	13	6	2	3	2
2	Persons	30	13	2	13	2
3	Total payments, thousand som	107	54	12	33	8
4	Average monthly income, som	298	348	500	214	315

Source: Survey Data

Of the additional sources of income, separate income from the private sector needs to be specified further. There are two types of income - income as salary as hired workers and income from independent entrepreneurial activity (see Table 5.8 on page 89). Around 9% (or 81 of 895 persons) of people of working age from the sample reported additional work in the private sector. The average salary was 37% higher than in the permanent work sector, but the average working period is normally approximately 5 months per year. The average hired labor cost is highest in Bakai-Ata Rayon and lowest in Talas Rayon. The majority of the hired laborers find temporary work as short-term seasonal agricultural workers (60.8%), or as construction workers (in the second place with 10.8%). The majority of workers who worked as hired laborers in the private sector is concentrated in Manas and Kara-Buura Rayon.

Almost two times less people earn an income as individual entrepreneurs. The distribution across the territory is wide, but again Kara-Buura Rayon demonstrates higher figures in this category. The average income is significantly higher in Bakai -Ata Rayon. Almost half (49%) of the people working as independent small entrepreneurs, have specialized in the transport services market. The second place in this category is occupied by the trade of food products (14%). In this category we also observe a higher volatility of the average income across different zones.

Table 5.8: Income from Private Sector Employment and Remittances

		Total sample	Talas	Bakai-Ata	Kara-Buura	Manas
I	Salary of hired workers					
1	Persons	81	12	11	27	31
2	Total income, million som	2.2	0.2	0.3	0.8	0.9
3	Average monthly income thousand som	6.1	5.0	6.8	6.3	6.1
II	Entrepreneurial income					
1	Persons	43	9	8	16	10
2	Total income, million som	2.6	0.3	0.8	1.2	0.3
3	Average monthly income thousand som	7.7	4.5	11.3	9.0	4.3
III	Remittances					
1	Households	24	4	8	4	8
2	Total transfers, million som	1.2	0.3	0.3	0.1	0.5
3	Average annual transfers thousand som	48	82	32	23	61

Source: Survey Data

The last of the observed sources of income is the remittances of labor migrants working abroad. In Talas Oblast, this activity is not as widespread as in South Kyrgyzstan, but a certain activity can be detected. The sample shows that 8% of the households receive additional support from former household members currently working abroad and transferred an annual amount of 1.2 million som. The overall sum of transfers is quite substantial, but the average sum fluctuates strongly, depending on the rayon. We cannot make a deeper generalization from the sample due to the low level of such an activity in the region. The only thing we could specify is the presence of this activity and its basic trends. Migration is equally distributed between Russia and Kazakhstan. The main sectors that employ labor migrants from the oblast are trade, the service sector (each 30%) and construction works (25%).

Thus, the total non-agricultural income and transfers received and generated by the rural households from different sources in Talas Oblast consist of 17.4 million som, or 40% of the total agricultural product. In a number of households, where an absence of land, livestock or labor force was observed, additional income sources complementarily support the well-being of the population and substantially affect the poverty rates.

6. Empirical Analysis Results

This chapter presents the final outcomes of the research project. The first section describes the poverty measurement procedure, including the method of defining aggregate consumption, poverty line adjustment and the results of poverty measurement in the area. The second section is dedicated to the measurement of the production efficiency indicators of the collected data and to the interpretation of the results. The final part of the chapter models poverty and production efficiency indicators and explains linkages of the variables and implications arising from the results.

6.1 Poverty Measurement

Following the developed practice of poverty measurement, we will finalize the process of poverty level assessment in the selected region in this section. It was specified earlier in Section 3.1, that the main steps in poverty measurement are identification and aggregation steps. The first step consists of selecting a unit of measurement (i.e., household), selecting and constructing a *resource* to measure (income or consumption) and defining and adjusting the poverty line. The second step is the application of poverty measures (indexes) to the collected and processed data sets. The first subsection explains the selection and construction of the measured resource - aggregate consumption on the household level. The following part is devoted to defining the poverty line and its adjustment. The final part of the section presents the results of the poverty measurement in Talas Oblast.

6.1.1 Consumption Aggregate Construction

The discussion on selecting the resource of poverty measurement between income and consumption is described in Section 3.1; it is standard practice to use aggregate consumption as the major indicator of the population's living standards in the developing countries, including Kyrgyzstan. The official approach to the consumption aggregate on the national level includes food and non-food products, services and other expenses as well as the imputed value of the purchased durable goods in the observed period (details see page 28). The alternative method of aggregate expenditures as analyzed by the NSC in 2002 also suggests, besides individual consumed goods and services, an extended range of durable goods, property and specific costs such as agricultural production costs, purchases and taxes.

There is no need to also discuss the inclusion in aggregate consumption of expenses on support to relatives and friends, costs related to celebrations and rituals, and neither the imputed value of durable goods and investments (purchase of agricultural animals, land as immovable property, and other types of durable goods). But additionally, production costs were added to this list, including veterinary services, purchase of seeds and fertilizers, and finally all types of taxes. However, it is stated in the literature that neither taxes, nor any business-related expenses can be part of the aggregate consumption (Deaton and Zaidi, 2002, p29). Therefore, we used the method of the personal consumption aggregate in the consumption aggregate's construction. The NSC informed us that this method has been used as the official approach since 2004, however it did not give any more precise information about the inclusion of such a controversial item. We suppose that the approach we applied here is identical to the official one.

Thus, the aggregate consumption expenditure of the rural household sample of Talas Oblast includes the following types of expenses:

1 purchased food products,

2 own produced food products consumption (crop and livestock products),

3 food products received for free as gift, support or donations,

4 non-food expenses,

5 expenses on medical treatment and medicines,

6 transportation expenses,

7 education,

8 celebrations, gifts and rituals,

9 support to relatives and friends,

10 communication expenses,

11 other expenses (details are given further),

12 durable goods:

      [a] automobiles,

      [b] land,

      [c] livestock purchase.

The full list of items, services and costs can be derived from the research questionnaire attached in Appendix A. Consumption costs data were taken from the questionnaire, but the method and type of information differs depending on the type of the consumption item. In different cases, different types of information and calculation schemes were applied.

Food products costs were measured in the first two categories, i.e. purchased food and own produced crop and livestock products. Purchased food includes main food and beverage items typical for Talas Oblast. The initial list of products was taken from two household questionnaires - the Fourth Living Standard Survey in Kyrgyzstan, arranged in 1997 (KPMS, 2002a, p92-96), and the latest version of the questionnaire used by KIHS (NSC, 2008, p4-24). Originally, it included 97 products typical for the Kyrgyz consumer market, but after the pilot stage the number of food items was reduced to 63 food items. Such a reduction became possible through excluding products not typical for the rural market or the consumption pattern in Talas Oblast and joining some similar products in one food or beverage item (e.g., tropical fruits, alcoholic drinks or tobacco products). Subsequently, purchased food expenditures costs may be defined as follows:

$$\mathbf{C}_{Food} = \sum_{i=1}^{63} \mathbf{P_i} \times \mathbf{Q_i} \times \mathbf{T}, \qquad (6.1.)$$

where $\mathbf{P}$ is the price of the ith (from 1st to 63rd position) food item, $\mathbf{Q}$ is the quantity of the ith food or beverage consumed, and $\mathbf{T}$ is the periodicity of food purchase (from daily to annual purchase). Each of the food items was calculated separately and then summarized. The prices of the purchased food were taken from the questionnaire and some deviation was observed across the sample. Price deviation is possible due to the difference of the source of purchase (private person, shop or market), periodicity (annual or weekly basis), and quantity of the purchase.

Own produced food products data were derived from the assessment of the agricultural production of the rural households included in the sample, presented in the previous chapter. The value of own produced crop and livestock products was calculated based on the data (quantity and price) taken from three sources - crop products produced in the main agricultural land plots, crop products grown in the kitchen gardens, and livestock product produced by the households.

Prices were not available for all products, therefore data for some products were taken from the data set of purchased products, using the same formula, 5.1., or were taken from the NSC database (explanation is provided on pages 78 and 83).

Products received for free are included in the consumption. This group of products includes all products received from relatives, friends or humanitarian organizations for the last 12 months. Average prices were not calculated, because besides the amount of the products the values of these donations are also given.

Non-food expenses include all types of expenses on the purchase of clothes, shoes, goods, textiles, electric appliances, furniture, jewelry, construction materials, agricultural tools and expenses related to automobiles. Originally, it was planned that expenses on furniture and electric appliances would be replaced by durable goods and analyzed separately, but only four records concerning these items, with a low average price, were found. Therefore, it was decided to keep this group of goods and expenses as it was. Only expenses for the last three months given for every listed item were summarized, without it being necessary to calculate prices.

Expenses on medical treatment and medicines include the basic costs incurred by household members for any type of medical treatment, medicines, analyses and tests, as well as costs of food and transport expenses. Data were collected on the basis of the expenses of the last quarter.

Transportation expenses were calculated as the costs of payments for different types of transportation (taxi, bus services and others) by all members of the households. Typically, the last quarter periodicity was applied.

Education includes costs related to the study process of different groups of children and young inhabitants - from kindergartens to higher education. Different types of expenses were listed - as direct costs (like the costs of contracts in colleges) and indirect costs, too (like sportswear and payments for the repair of schools). Data include the actual costs during the last academic year.

Expenses on celebrations, gifts and rituals are part of the traditional customs of Kyrgyz culture. 68% of households report about these expenses. This type of expenses is often of an extremely high value in the case of funerals or weddings. Even if a household does not arrange a public funeral or wedding, the custom of presenting high-cost presents, livestock or simply money typically exists among relatives within the clan system. This phenomenon still needs to be researched in detail from the point of view of behavioral economics. It is clear that a certain network insurance system was created under the conditions of the weakening role of the social protection system in Kyrgyzstan, but only a deeper study may suggest the ways of a positive transformation of the developed celebration customs. The aggregated sum of the annual expenses on these purposes is given. It is not clear from the explanation of the methodology (NSC, 2011c, p1-4), whether they include the personal consumption expenses on celebrations, gifts and rituals or not. We suppose that due to the high influence of these expenses it is necessary to include them in the consumption.

Support of relatives and friends normally includes the help provided for close relatives or friends or even neighbors in form of food products, money, clothes, etc. Usually, support is provided by parents to children or brothers and sisters. This support is different from previous types of support in the social form of presenting. In case of celebrations and gifts it is always a public act of presenting, usually arranged during a certain period of the year (seasons of celebrations, weddings always start after harvest). Support is arranged in more private situations and is understood as a person-to-person relation. Annual base expenses are included in these expenses. In case of presenting food products, average prices are taken from the data set of own produced food.

Communication expenses consist of the costs of mobile and stationary telephone communications. Here, we need to stress that the mobile sphere has developed quite fast in the country within the last decade and is affordable for most of the country's population. 94% of the respondents report about costs for mobile communications, while only 2 of the covered 297 households informed us about the availability of stationary telephones. Expenses were given for the last quarter.

Other expenses include expenses for the purchase of heating materials, electricity, water, garbage disposal, aliments, food out of home, costs of own processed food (jams, juices, marinaded vegetables, etc.), tourism and resort expenses, law services, insurance costs. Data were usually given for the last quarter, except for water and electricity, which were given on a monthly basis, and for heating materials and own processed food, for which annual quantities were provided.

Measuring durable goods' imputed value for aggregate consumption requires the definition of several important factors, i.e. the current value of the durable goods, the real rate of interest and the rate of depreciation for durable goods.

The last two rates are hard to evaluate under Kyrgyz conditions, due to the absence of adequate methods of defining the depreciation rate on the one hand and high market interest rates on the other. Therefore, it was suggested in the literature, specifically for Kyrgyzstan, to use a simplified assumption of a unified rate of 10% applied to the current value of all durable goods (Deaton and Zaidi, 2002, p34).

The original LSMS questionnaire includes a full list of property items which belong to a household; besides the name and quantity it also consists of columns concerning the year of purchase and the purchase price.

At the pilot stage it was found that collecting this information fully is one of the most time-consuming procedures and additionally requires a lot of patience from the respondents. Due to a serious restriction in time and resources, this information collection was substantially reduced. Therefore, full information on the values of durable goods is not available to us.

Nevertheless, we were able to define such basic durable goods as automobiles, and also the purchase of land in the last three years, as well as the purchase of livestock.

Almost one third of the households in the sample had their own automobiles. Thirteen households had two automobiles, and eighty two owned at least one automobile. From total 108 automobiles 90 were personal automobiles. Besides that in the motor-vehicle pool structure were also included 14 lorries and five minibuses. Most of the automobiles were quite old, with the average production year 1990, and the average year of purchase 2006. For each automobile, we asked for the current estimated market value. From this sum a suggested 10% of imputed value was included as expenses on durable goods.

Also, we collected data concerning the purchase of land plots over the last three years. There were only five cases of purchases of 17 hectares of land for a total amount of 1.1 million som. The same 10% depreciation rate was applied here, too. It is not clear how the aggregate consumption value of purchased livestock was counted in the official estimations. We suggest to use the calculated imputed value of the animals (similar to the depreciation rate). We decided to limit livestock types included in the durable goods to the main productive animals - cattle, sheep and horses. From the collected data we learned the share of animals sold each year. For cows and horses this share is 15%, for sheep it is significantly higher and reaches 25%. We assume that farmers sell animals to save a stable productive livestock in the productive age range. We used these rates to evaluate the 'depreciation' value of the purchased livestock.

In relation to outliers, the literature suggests to check for possible misunderstandings of the codes and for misinterpretation of the questions (Deaton and Zaidi, 2002, p23). However, the quality of the data was high enough to detect a minimal number of such types of mistakes. Another option was the use of the adjustment procedure, but in the Kyrgyz poverty survey documentation this procedure was only mentioned once (Ackland, 1996, p34-35), while it was not mentioned in later surveys. Therefore, we did not implement this procedure for an adjustment of aggregate consumption outliers. Also, costs related to credits, debts, and the purchase of financial services were not included in the consumption estimates (Deaton and Zaidi, 2002, p30).

The structure and consumption pattern of the sample and the distribution by rayons is shown in Table 6.1 on page 97. The structure shows large differences in consumption across the territory. In Manas Rayon, the share of purchased food was higher then the area's average, while the share of own produced food is the lowest in the area. The share of electricity expenses is higher in the most mountainous part of Talas Oblast in Talas Rayon. Another difference in the consumption pattern that needs to be specified separately concerns the higher expenses on education in Kara-Buura and Bakai-Ata Rayon. The difference arises due to higher costs on special secondary and higher education.

Table 6.1: Structure of the Consumption Aggregates Across the Target Region

	Consumption categories	Total	Rayons			
			Talas	Bakai-Ata	Kara-Buura	Manas
1	Purchased food	39%	37%	38%	38%	43%
2	Own produced food products	15%	15%	16%	18%	12%
3	Food given for free	0.2%	0.4%	0.1%	0.2%	0.3%
4	Non-food expenses	11%	12%	11%	10%	11%
5	Expenses on heating materials	3%	2%	3%	4%	5%
6	Electricity	4%	5%	4%	2%	3%
7	Medical treatment and medicines	4%	4%	3%	4%	4%
8	Transport expenses	2%	3%	2%	2%	2%
9	Education	3%	2%	3%	4%	1%
10	Celebrations, gifts and rituals	9%	8%	9%	9%	9%
11	Support to relatives and friends	2%	2%	3%	1%	0.5%
12	Communication	2%	2%	2%	2%	2%
13	Other expenses	3%	3%	3%	4%	3%
14	Durable goods value:					
a	Purchase of animals	1%	2%	1%	1%	1%
b	Automobiles	3%	3%	3%	3%	3%
c	Land	0.2%	0.3%	0.2%	0.4%	0.0%
I	Aggregate Consumption Total, million som	49.3	10.7	14.3	13.8	10.5
II	Aggregate Consumption per HH, thousand som	166.2	153.4	183.2	181.5	144.3

Source: Survey Data

6.1.2   Poverty Line Adoption

The poverty line applied in Kyrgyzstan is an absolute type of poverty line devoted to the detection of the share of people, who are most in need, in an attempt to afford the basic needs. The poverty line was adjusted several times in the 1990s. Since 2003 the methodology has not changed anymore and is calculated by the NSC every 5 years (calculation methods are explained on page 27). The last recalculation was performed in 2008 (see Table 3.1 on p. 27). Since then, the poverty line is adjusted annually based on the CPI (Consumer Price Index). The last reindexation was performed in 2010 and poverty lines were defined as follows (NSC, 2011c, p18-19):

- absolute poverty line - 20 937 som per capita,

- extreme poverty line - 12 608 som per capita.

Research was conducted in the period of May-June 2011. Therefore, we used the simple application of the CPI growth index, from December 2010 to June 2011 as the poverty line deflator. This CPI index equated to 107.2% for the Kyrgyz Republic and 110.3% for Talas Oblast (NSC, 2011b, p26). Subsequently, we multiplied the available poverty lines by a CPI index of Talas Oblast and received the following poverty levels:

- absolute poverty line - 23 093 som per capita,

- extreme poverty line - 13 907 som per capita.

These figures are robust enough, because we do not precisely know the detailed technique used by the NSC. But we believe that a more precise way may improve the final figures by some percent only.

The next important issue when applying the household consumption aggregate to the poverty line is the necessity of adjusting the household consumption by adult equivalent to take into account the household size and composition influence. Literature discussed the necessity of an adjustment of consumption, based on the argumentation that an economy of scale exists in big families and adult consumption tends to be higher than children's consumption. A number of approaches were developed for this purpose (Deaton and Zaidi, 2002, p46-52). However, the per capita consumption approach was applied in Kyrgyzstan, because of certain problems related to the application of the adult equivalent scales. It was noticed that at any scale such adjustments simply bring about a lowering of the poverty assessment, specifically for families with a large number of children. Another argument was that in Kyrgyzstan food consumption is critically important for the composition of the consumption basket, and an economy of scale argument cannot be applied for food (WB, 2007a, p11). Thus the per capita consumption was selected as the appropriate choice.

6.1.3 Poverty Measurement Results

Talas Oblast demonstrates a high volatility of the poverty trends over the last five years. Improvements were not sustainable and changes regularly caused strong declines (see Table 6.2). It is interesting that urban poverty in Talas Oblast creates a certain gap in contrast with rural poverty. Since 2008, even after the decrease, urban well-being has demonstrated the higher level. Unfortunately we cannot analyze all of the poverty indexes for Talas Oblast. However, it is quite symptomatical that the high volatility of the poverty trends coincided with the change of the Poverty Gap and Poverty Severity Indexes on the country level (5*th* and 6*th* columns of Table 6.2). It means that even poverty trends stabilized on a certain level on the country level between 2007 and 2010 (see Figures 2.3 and 2.4 on pages 12 and 13). Controversial trends among the poor strata of the population in the region additionally reflected the complementary poverty indexes and need to be further analyzed.

Table 6.2: Poverty Indexes in Dynamics in Talas Oblast and in Kyrgyzstan in 2006-2010

	Talas Oblast				Kyrgyzstan	
	Head Count Index				Poverty	Poverty
	Absolute Poverty,%	Absolute Urban,%	Absolute Rural,%	Extreme Poverty,%	Gap Index,%	Severity Index,%
2006	40	42.2	40	9.7	9.1	3.1
2007	35.3	36	35.3	7.9	6.6	1.9
2008	43	38	43.9	4.6	7.5	2.6
2009	33	24.9	34.4	2.9	5.1	1.8
2010	42.3	34.3	43.7	-	7.5	2.5

Source: Poverty Reports (WB, 2007a; NSC, 2011c

Based on the survey data and methods described in earlier subsections we defined an alternative poverty picture of the rural areas of Talas Oblast for the middle of 2011. We calculated the standard poverty indexes - Headcount Index (HCI) - P_0, Poverty Gap Index (PGI) - P_1 and Poverty Severity Index (PSI) - P_2 on the oblast and rayon levels. For the HCI two poverty rates were calculated according to two poverty lines - the absolute and the extreme. The results are presented in Figure 6.1 on page 100.

Looking at the absolute poverty level of Talas Oblast's rural population, the HCI has placed 39% of the population below the absolute poverty line (or 31.6% of the rural households). The extreme HCI shows a poverty rate of 5.6% (or 4.3% of the households). Across the rayons, the poverty rate supports the aggregate consumption pattern in relation to the absolute poverty line, but in case of extreme poverty, the assessment did not concur. This simply means that poor people concentration in different parts of the region different.

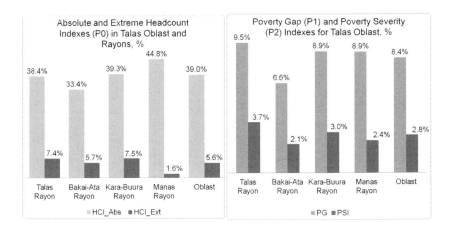

Figure 6.1: Poverty Indexes in Talas Oblast and Rayons
Source: Author's own calculations from Survey Data

Some more information may be presented by the right slide of Figure 6.1. Poverty Gap (PGI) and Poverty Severity Index (PSI) support the conclusion that the best situation on all levels with regard to poverty can be found in Bakai-Ata Rayon. In other rayons, the situation is more complicated.

According to the Absolute HCI, Talas Rayon shows the second lowest poverty level, but the Extreme HCI is higher than the average level in the oblast. The Poverty Gap Index tells us that the mean consumption of the poor in Talas Rayon is lower in comparison with Bakai-Ata Rayon by 2.9% with regard to the poverty line. Thus, in order to alleviate poverty, it would require about 44.5% more money to overcome the gap to the poverty line in Talas Rayon than in Bakai-Ata Rayon, if all other factors were equal.

The Poverty Severity Index in Talas Rayon shows us that the share of poor people with regard to their extremely lower mean consumption is higher by 74% in comparison with Bakai-Ata Rayon. Finally, this results in the fact that 7.4% of the people live below the extreme poverty line, as opposed to 5.7% in Bakai-Ata Rayon. Thus, the absolute poverty rate often hides the incidence and depth of the poverty situation.

Similarly, we could argue that Manas Rayon shows the highest poverty rate. The HCI is equal to 44.8% of the population. The PGI is also close to the level of Talas Rayon. Nevertheless, the PSI is much better here (2.4%) and as a result the low level of extreme HCI is the best in the region (1.6% or one household only in Manas Rayon is below extreme poverty line). This means that despite a higher poverty rate in relation to the absolute poverty line, the majority of households is located not far from the mean well-being, and income distribution is more equal than in other rayons. Deviations from the mean consumption in Manas Rayon are rare and mainly located around 9% below the absolute poverty line, while in other rayons people live more unequally distributed below the poverty line.

In general, the poverty situation in Talas Oblast repeats the trends which are typical for the country's rural population. Partly because of the per capita approach, described previously, poverty incidence strongly depends on household size (WB, 2007a, p11). Other interesting specifics were presented to us through the distribution of poverty in household head sub-groups by gender and age (see Table 6.3).

Table 6.3: Poverty Indexes by Household Head Gender and Age Sub-Groups

	HCI Absolute, %	HCI Extreme, %	PG, %	PSI, %	Population Share, %	HH size, persons
Household Head Gender						
Male	39.9	6.0	8.9	3.0	86.0	5.7
Female	33.6	3.5	6.1	1.8	14.0	4.3
Household Head Age						
19-25	33.3	22.2	11.4	4.0	1.7	4.50
26-34	56.9	0.0	12.3	3.5	8.9	5.33
35-44	45.2	5.9	10.4	3.6	26.4	5.62
45-54	30.6	4.3	5.8	2.0	27.5	5.24
55-64	20.9	11.5	8.1	2.8	22.2	5.77
65+	34.4	0.0	7.6	2.3	13.3	5.24

Source: Calculations from Survey Data

It might be easy to notice that households headed by females demonstrate better poverty indicators on all levels. But it has to be pointed out that this small group of households is mainly headed by widows of a high average age (see page 73). This fact in turn leads to a smaller size of the households in this sub-group and potentially to a higher additional income from pensions.

An analysis of the household head age sub-groups shows that the best performance is demonstrated by heads from 45 to 54 years of age. That may be explained with a certain lifetime cycle of the household. The starting position of the household (from 19 to 25 years) is provided with the support from parents and poverty incidence is not strong. However, deviations from mean values are high due to different starting conditions. The age from 26 to 34 years is the hardest period in the family formation, coinciding with the appearance of children and low income capacities. Starting from 35 years, the situation gradually improves and when the head reaches the age of 45, children are transferred to the working age group and the household gains from the additional work force support. A certain volatility in the period from ages 55 to 64 is characterized by high male mortality among the heads of households and the necessity of reforming the household (creating new families). The next age period is characterized by a certain stability of well-being. This picture may not be absolutely correct, but in general it is also repeated on the country level (NSC, 2011c, p22-23).

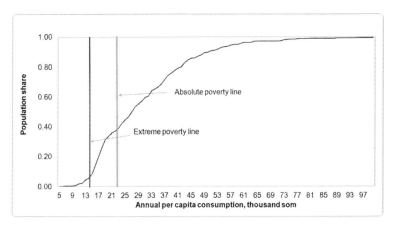

Figure 6.2: Cumulative Distribution Function of Consumption of Rural Population Consumption in Talas Oblast
Source: Survey Data

The poverty incidence curve in Figure 6.2 demonstrates that sensitivity to the changes of the extreme and absolute poverty lines will be different, depending on current consumption distribution. Certainly, an adjustment of the poverty line in case of inflation may bring about a serious volatility of the poverty indexes in the future.

The extreme poverty index is more sensitive to such deviations due to the changing slope of the curve. The poverty indexes in Talas Oblast fit in the overall picture of the poverty trends in Kyrgyzstan and in the region. Poverty deviations are detected across the region, however they coincide with the descriptive data observed in the previous chapter.

A considerable influence on poverty incidence is provided by traditional demographic parameters - household size, head gender and age. In general, poverty in the region is not very deep and chances of potential improvements are high. Absolute and extreme poverty indexes are not very informative with regard to poverty depth, and PGI and PSI indexes support a better analysis of the poverty distribution below the poverty line.

6.2 Production Efficiency Measurement

This section describes the data, methods and results of the production efficiency measurements on the rural household level. Two different approaches were applied to derive the production efficiency of the rural household - the Data Envelopment Analysis (DEA) and the Stochastic Frontier Analysis (SFA).

6.2.1 Production Variables Set

The main production features of household production were presented previously in Chapter 5. However, we additionally need to specify the set of the basic variables used to define production efficiency estimates on the household level. These variables are identical in the DEA and SFA approaches - output and input variables.

The output variable is the aggregated agricultural production of the household. As we already know, it mainly consists of the different crop and livestock products, and is defined as follows:

$$Y_i = \sum (CP_i + LP_i + LR_i), \qquad (6.2.)$$

where, Y_i is the agricultural output of the ith household; CP_i is the crop production of the ith household (production of the main land plot and kitchen garden, including sold, stock and consumption values); LP_i is the production of the livestock products of the ith household (livestock and livestock products sale, livestock products consumption and livestock capital value increase); and LR_i is the income of the ith household from renting out its land plot (details see Table 5.4 on page 85).

The labor variable consists of the number of household members in the working age of 15 to 65 years per household. Official statistics count the people starting from 17 years of age, but due to practical observations made in the course of the field study over the summer period, children are involved in field work from an even earlier age. People receiving disability pensions were excluded from the number of people in the working age. Thus, 27 persons with different disabilities from 26 households were not counted as workforce. All of them were of working age.

The capital variable was calculated as the sum of the two main production factors - land and livestock. The land factor itself consists of two different land types - the main land plot (including all types of production land) and the kitchen garden. In both cases, land prices were self-estimated, but usually the kitchen garden was evaluated at a higher price than the field plot price. The land price was defined as follows:

$$LC_i = S_i \times ALP, \tag{6.3.}$$

where LC_i is the land capital of the ith household in thousand som, S_i is the acreage of the land plot in hectares, ALP is the average land price in the sample, thousand som per hectare. ALP was calculated by the average price formula 5.1. (see page 77), but instead of the actual price, respondents were asked for the price they might sell their land plot for. Land capital was calculated separately for the main land plot and the kitchen garden. The average land price for the main land plot amounted to 80.4 thousand som per hectare, while the kitchen garden was evaluated at 234.4 thousand som per hectare. The kitchen garden was evaluated at a more expensive level, because typically it is linked to the house, easily accessible and more productive.

The livestock capital was calculated as the market price of the livestock belonging to the household. Three main types of livestock were included in the assessment of production livestock - cattle, sheep and horses. The price for the livestock was defined separately for each household. The difference between livestock and land lies in the market character of livestock, which can be easily converted into cash and back again. For instance, 23% of all livestock is sold annually and converted into monetary value, while sales or purchase are quite rare in the case of land. Thus, livestock prices are more precise and realistic.

289 households (97%) of the sampled 297 households demonstrated positive non-zero values for all variables - agricultural output, labor and capital factors. The 8 remaining households (or the remaining 3% of the sample) were not included in the production efficiency assessment. Only one household in this group demonstrated high production activity, although it consisted of one person only who reported to be the recipient of a disability pension. In other cases these non-agricultural households mainly showed alternative income sources - income from remittances, pensions, the service sector or official employment. The average household size of non-agriculturally oriented households is more than two times lower than average in the target area i.e. 2.5 persons per household; the average age is also significantly higher i.e. 48.6 years. These non-agricultural households usually consist of adult persons - pensioners, persons with an income from commercial activity or permanent work, typically without children.

Of the selected 289 households, approximately 7% did not own a land plot and livestock, and produced an insignificant amount of kitchen garden products (around 14 thousand som

Table 6.4: Descriptive Statistics of the Production Variables of the Sampled Households

	Agricultural Output, Thousand Som	Labor, Persons	Capital, Thousand Som
Mean	147.9	3.2	348.4
Median	103.8	3.0	248.3
Maximum	2 135.9	7.0	10 428.6
Minimum	1.5	1.0	14.1
Std. Dev.	176.6	1.4	657.4
Sum	42 737	926	100 696

Source: Calculations from Survey Data

per household). Such a micro-scale production certainly increases the error terms of the proposed model, but still it was decided to keep these households in the data set to analyze all levels of production scales.

6.2.2 DEA Efficiency Model Application

Among DEA models, the input-oriented constant returns to scale (CRS) and variable returns to scale (VRS) models introduced, respectively, by Charnes et al. (1978) and Banker et al. (1984) are used most often (Coelli et al., 2005, p162). Both models present deterministic non-parametric efficiency estimations of the rural households in the sample of Talas Oblast with different scale assumptions.

The DEA model efficiency was calculated with the specialized software tool "Data Envelopment Analysis (Computer) Program" (DEAP), version 2.1. This free software was developed by the Australian production economist, Prof. Dr. T.J Coelli (Coelli, 1996a, p1-50). DEAP 2.1 allows the estimation of a number of DEA production efficiency models, including CRS and VRS DEA models, cost and allocative efficiency models, and Malmquist DEA methods for panel data. For our purposes we needed to calculate technical and scale production efficiencies (Färe et al., 1994, p34-37,49-53) in the CRS and VRS model options of the software.

The input-oriented DEA CRS consists of several stages of solving two stage linear program problems - definition of the initially solved ratio-based efficiency measure and at the second stage, problems of slacks digression. The core description of the methods used by the software was derived from Coelli (1996a, p1-50) and Coelli et al. (2005, p162-167).

The model was initially constructed for a set of given individual data, a non-parametric envelop frontier covering all the producer's output points which lie below the frontier or exactly on it. We propose to define the existence of I farm-households (further farms), producing A outputs by using B inputs. A specifically selected i^{th} farm could be represented by vectors y_i for outputs and x_i for inputs.

Two matrixes were created for output and input data sets. The Y output matrix consisted of $A \times I$ vectors, while X was represented by $B \times I$ input vectors for I farms. For each of the farms the ratio's measure of all output over all inputs was defined as $u'y_i/v'x_i$, with u representing the vector of output weights, and v the input vector weights. Mathematically, the Linear Programming (LP) problem was solved by:

$$\max_{uv}(u'y_i/v'x_i),$$
$$\text{subject to} \quad (u'y_j/v'x_j) \le 1 \qquad j = 1,2,\ldots,I, \qquad (6.4.)$$
$$u, v \ge 0$$

Thus, the values of u and v were defined to maximize the ith farm efficiency, with the bounding limits of efficiency between 0 and 1.

The next problem that arose due to this approach was the unlimited number of solutions, which were solved through imposing a constraint $v'x_i = 1$, which defined the following solution:

$$\max_{\mu v}(\mu'y_i),$$
$$\text{subject to} \quad \nu'x_i = 1$$
$$\mu'y_j - \nu'x_i \le 0 \quad j = 1, 2, \ldots, I, \qquad (6.5.)$$
$$\mu, \nu \ge 0,$$

A change of the notations from u and v to μ and ν was used to solve them as different LP problems. The form of LP problem defined in 6.5. is known as multiplier form following (Coelli et al., 2005, p163).

The envelopment form of the problem is derived as follows:

$$\min_{\theta\lambda}\theta,$$
$$\text{subject to} \quad -y_i + Y\lambda \ge 0,$$
$$\theta x_i - X\lambda \ge 0, \qquad (6.6.)$$
$$\lambda \ge 0,$$

where θ is a scalar measure and λ is a $I \times 1$ constant vector. This solution of envelopment is preferable due to lesser constraints than those presented by the multiplier form ($A + B < I + 1$). θ presents the efficiency rate for the ith farm. θ is limited by the value of 1 and the means of the frontier, and defines a technically efficient farm. LP problem solves for each of the farm in the sample and received θ values are efficiency scores.

In the literature, Färe et al. (1994) and Coelli et al. (2005, p164), production technology as presented by system 6.6. is defined as $T = \{(x, y) : y \leq Y\lambda, x \geq X\lambda\}$. The production set is closed and convex, and demonstrates CRS and a strong disposability level.

The second stage LP problem of slacks was graphically presented in the theoretical part (see Fig. 3.1 on page 39). We only need to refer to the slacks problem again to describe the solution of the problem. For defined optimized values of θ and λ, the measured input slacks $IS = 0$ if $\theta x_i - X\lambda = 0$. The problem may be solved as follows:

$$\min_{\lambda, OS, IS} -(A1'OS + B1'IS),$$
$$\text{subject to} \qquad -y_i + Y\lambda - OS = 0,$$
$$\theta x_i - X\lambda - IS = 0, \qquad (6.7.)$$
$$\lambda \geq 0, OS \geq 0, IS \geq 0,$$

where OS presents the vector of output slacks $(A \times 1)$, IS defines the input slacks vector $(B \times 1)$ while $A1$ is a $A \times 1$ vector and $B1$ is a vector of $B \times 1$.

The slacks problem was solved for each of the farms and the θ value was derived from the first stage LP used here. DEAP used the multi-stage procedure by conducting a sequence of radial linear programs to identify the efficient projected points (Coelli, 1996a, p15). In our analysis we only used the LP solution described in program 6.6., which gave us the primer technical efficiency measure.

The rationale of skipping slacks is based on the argument that they are a result of limited samples, because if we increase a sample to extremely large or infinite values, or change the frontier construction from piece-wise to smooth construction than slacks will disappear from the calculations (Coelli, 1996a, p15). A more practical reason is the unclear interpretation of the slacks analysis on the further application to poverty issues, as the decrease of slacks in inputs cannot be utilized optimally in the remote rural areas. The lack of non-farm employment for labor input, or critical limitations on the land market, deteriorate the potential for an optimal use of inputs, except for the livestock factor.

The input-oriented VRS model was developed as a continuation of the CRS model to overpass critical assumption of the CRS model, supposing that all producers operate on an optimal level. However, numerous reasons often bring producers to a non-optimal production level. Thus, VRS is simply an extension of the CRS model, which accounts for VRS cases (Coelli et al., 2005, p172). In this model, the technical efficiency is complemented by the measure of scale efficiency. The problem is solved by introducing the convexity constraint condition

$(I1'\lambda = 1)$ to LP 6.6., which is finally defined as follows:

$$\min_{\theta\lambda}\theta,$$
$$\text{subject to} \quad -y_i + Y\lambda \geq 0,$$
$$\theta x_i - X\lambda \geq 0, \qquad\qquad (6.8.)$$
$$I1'\lambda = 1$$
$$\lambda \geq 0,$$

where $I1$ is an $I \times 1$ vector of the production units. Technical efficiency defined within this model could be decomposed into technical efficiency and scale efficiency by conducting the Data Envelopment Analysis program of CRS and VRS to the same set of data. The difference in results of technical efficiency between CRS and VRS indicates the actual value of the scale efficiency.

Figure 6.3 presents the graphical illustration of the scale efficiency for the case of a one input - one output production unit. Two different frontiers are presented here - for CRS and VRS cases. In the case of CRS input, the oriented technical efficiency for farm P will be equal to the distance PP_C, but for VRS it will be presented as PP_V. The difference between these points $P_C P_V$ is the scale efficiency. The solution of this problem in ratio efficiency measures is presented as follows:

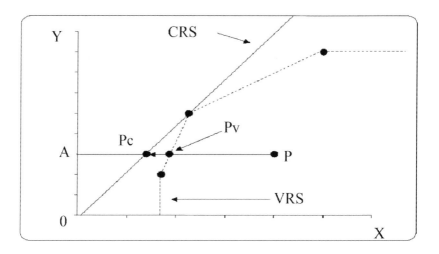

Figure 6.3: Calculation of Scale Efficiencies in DEA
Source: Author's own representation following Coelli (1996a)

$$TE_{I,CRS} = AP_c/AP,$$
$$TE_{I,VRS} = AP_v/AP,$$
$$SE_I = AP_c/AP_v, \qquad \text{while,}$$
$$0 < TE_{I,CRS} < 1,$$
$$0 < TE_{I,VRS} < 1, \qquad \text{(6.9.)}$$
$$0 < SE_I < 1,$$

and

$$TE_{I,CRS} = TE_{I,VRS} \times SE_I, \qquad \text{due to,}$$
$$AP_c/AP = (AP_v/AP) \times (AP_c/AP_v)$$

For our purposes we used both of the models to receive technical and scale efficiency estimates for the sampled households. For input-oriented constant returns to scale, the DEA model used the one output - two inputs scheme. As the dependent variable we used the aggregated agricultural production and as explanatory variables the capital and labor values. The second DEA model concerning input-oriented variable returns to scale used a one input - one output framework. Therefore, two different efficiency estimations models, based on the VRS Scale DEA model for two main inputs (capital and labor), were conducted separately here. Due to the method of efficiency estimation, each of the VRS models produced three estimations - at CRS, at VRS and Scale Efficiency. Statistics of the mentioned models' efficiency estimation are presented in Table 6.5.

Table 6.5: Sample Descriptive Statistics of the DEA Efficiency Estimations

	Mean	Median	Maximum	Minimum	Std. Dev.
$DEA1_{CRS}$	0.37	0.31	1	0.01	0.22
$DEA2_{CRSCAP}$	0.21	0.19	1	0.01	0.13
$DEA2_{VRSCAP}$	0.30	0.23	1	0.04	0.21
$DEA2_{SCALECAP}$	0.81	0.94	1	0.05	0.24
$DEA2_{CRSLAB}$	0.09	0.07	1	0.00	0.10
$DEA2_{VRSLAB}$	0.40	0.33	1	0.14	0.21
$DEA2_{SCALELAB}$	0.23	0.19	1	0.00	0.17

Source: Calculations from Survey Data

The DEA models at CRS and VRS demonstrate different efficiency estimations. In general, the CRS model shows quite a low average efficiency - around 40% from the maximum level. The distribution of households reveals that a relatively small share of households show high and maximal efficiency estimates (see Figure 6.4). The majority of households shows an efficiency lower than 50% from the frontier. Such a distribution supports the theoretical expectation from the model, supposing a small amount of efficient producers,

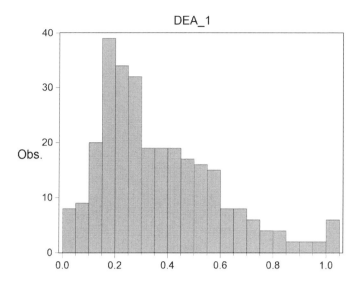

Figure 6.4: Distribution of DEA Input-Oriented CRS Efficiency
Source:Survey Data

and low average productivity. It may be explained by the extremely low-scale character of production sites, which presumably cannot demonstrate high efficiency estimates at proposed estimations.

The VRS scale efficiency model brought different results. Figure 6.5 presents the stage-by-stage efficiency estimation distribution of the sample for capital-oriented scale efficiency. The shapes of efficiency distribution under CRS and VRS conditions demonstrate a high level of similarity to each other. This results in a higher scale efficiency level than in the case of the previous model. This may be explained by the fact that rural households reach at most a certain limit of the exploitation of capital resources, in our case of land and livestock. Another argument for this may be a low scale of the production size. In this case, the low difference between CRS and VRS means that the frontier for both return to scale cases is the same in most of the observations, and a change of the return to scale condition does not lead to an improvement of the scale efficiency estimation. Scale efficiency is highly realized by the rural household, but we cannot draw a definite conclusion between equally possible explanations.

At the same time, average VRS efficiency is equal to 30%, and scale efficiency exceeds 80% of the maximum potential production. The overall low level of efficiency at CRS and VRS in combination with high scale efficiency demonstrates that rural households cannot use

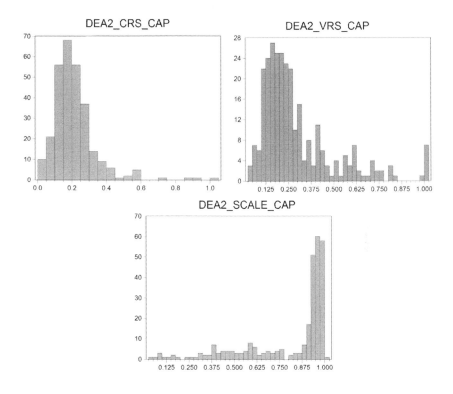

Figure 6.5: Distribution of DEA Input - Oriented Scale Efficiency (Output-Capital)
Source:Survey Data

typically the capital factor more intensively at the current level of technology and available capital (land and livestock) concentration. To reach a higher level of the economy of scale in agricultural production is quite problematic at the current level of capital concentration, while technology still might be improved even at the small-scale production.

The next VRS model, presented in Figure 6.6, shows that the capital efficiency in comparison with the labor force efficiency estimation is quite the opposite. Under CRS conditions, the labor force demonstrates an extremely low efficiency (9% at average), while the introduction of VRS conditions improves efficiency estimates (40%). It means that at CRS the distance between the extreme maximum outliers and average efficiency reaches big values, but that the introduction of VRS helps to rearrange the efficiency estimates of the labor scale efficiency. On the middle slide in the graph, we are able to recognize several peaks of concentration of households at the level of 25%, 35% and 50% efficiency. A small group of households

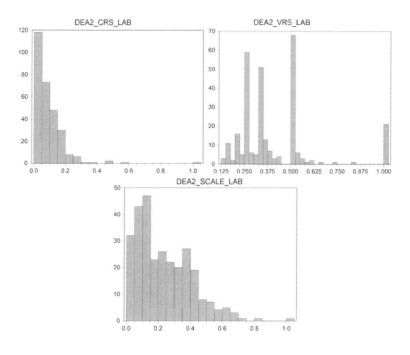

Figure 6.6: Distribution of DEA Input - Oriented Scale Efficiency (Output-Labor)
Source:Survey Data

demonstrates an extreme level of labor force efficiency. On average, scale efficiency reaches 23% when introducing variable returns to scale between CRS and VRS. The significant sensitiveness of introducing VRS to the labor-oriented DEA model may be explained by the structure of the labor force concentration per household. The change of the return to scale condition shows several points where labor productivity may be realized. The slope of the frontier, which envelopes data at the time of introducing VRS, changes significantly. Comparing the effect of introducing scale efficiency by estimating efficiency at different return to scale conditions between capital and labor we could conclude that the capital production limit is overexploited in comparison with the inefficiency of the labor force, which demonstrates a high potential for improvement.

It is useful to analyze the disaggregation of the sample into capital scale efficient and labor scale efficient households. In Figure 6.7, all the households are plotted across two efficiency scales estimations - labor and capital. Those households, which show a high efficiency using the capital factor, demonstrate quite a low or moderate labor scale efficiency, while labor

112

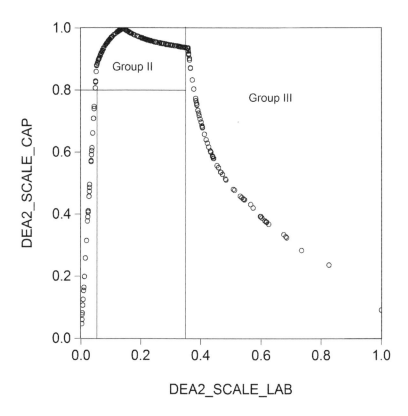

Figure 6.7: Distribution of Labor-Capital DEA Input-Oriented Scale Efficiencies
Source:Survey Data Calculations

efficient households demonstrate gradually decreasing efficiency from the capital point of view. One of the conclusions that can be drawn is that three group of households can be defined based on this presentation. Those groups are labeled in Figure 6.7. Group I consists of households that simultaneously demonstrate capital efficiency values of lower than 88% and labor efficiency of less than 5%. This small group (11% of the sample or 33 households of the 289 included in the modeling) is characterized by its low labor force (2.3 workers per HH on average) and capital concentration (80 thousand som per HH on average) and most likely depends on a non-farm income. Group II includes households with a capital efficiency of 88% to 100% and a labor scale efficiency between 5% and 35%. This group is the core of the sample (65% - 188 HH) and is characterized by higher labor (3.2 worker) and capital values (243 thousand som). The importance of labor and capital efficiency in those households is mixed. Group III includes households with a labor efficiency which is higher than 36% and a decreasing capital efficiency from 93% to 9%. This group consists of the rest of the sample (24% - 69 HH) and is characterized by significantly higher capital values (756 thousand som) and a bigger labor force (3.6 workers). Thus, significant concentration of the capital does not lead directly to a proportional increase of the agricultural output, but substitute the efficiency of the labor factor by capital. We may assume that the rural households select different strategies based on the availability of capital or labor concentration. The efficiency of the richer households of Group III shows that a decrease of capital efficiency accompanied by simultaneous increasing the labor efficiency. Poorer households of Group II combine the overuse of the capital factor with the average labor efficiency, while households of Group I cannot successfully use the production factor due its scarcity, and probably compensate it through alternative sources of income.

Of the DEA efficiency measures received in the poverty-efficiency linkage stage, all defined efficiency indicators will be used: of the CRS DEA technical efficiency estimations (one output - two inputs model) - $DEA1_{CRS}$, of VRS DEA (one output - one input model) technical efficiency and scale efficiency estimators - $DEA2_{CRSCAP}$, $DEA2_{VRSCAP}$, $DEA2_{SCCAP}$, $DEA2_{CRSLAB}$, $DEA2_{VRSLAB}$, $DEA2_{SCALELAB}$.

6.2.3   SFA Efficiency Model Application

The estimation of the Stochastic Frontier Analysis model efficiency was performed using the software package FRONTIER 4.1, developed by Coelli (1996b, p1-33). It was specifically developed to estimate the various types of SFA production efficiency models, including panel data. We were interested in the firm specific estimations of the so-called simple CD production function, using cross-section data. The software was distributed by the author free of charge, similar to the DEA software package.

For the application of the parametric SFA we suggest to use a simple Cobb-Douglas (CD) production function, based on the cross-sectional variables set and assuming a half normal form of distribution of the error term (for details see Fig. 3.3 on page 42). The initial proposed specification of the model was defined in formula 3.17. on page 39. The proposed CD production specification can be presented in the following form:

$$\ln Y_i = \beta_0 + \beta_1 \ln K_i + \beta_2 \ln L_i + \nu_i - u_i, \qquad (6.10.)$$

where Y_i is the output production of the ith household, K_i is the capital factor, and L_i is the labor force. ν_i and u_i are the noise component and the technical inefficiency component of the i-th household's error term, respectively, while $i = 1, \ldots, 289$. The properties of ν_i and u_i were defined earlier in assumptions 3.25. and 3.26. on page 42.

The parametrization of Battese and Corra (1977), which is used here suggests to substitute σ_ν^2 and σ_u^2 by $\sigma^2 = \sigma_\nu^2 + \sigma_u^2$ and $\gamma = \sigma_u^2/\sigma_\nu^2 + \sigma_u^2$ (Coelli, 1996b, p5). This procedure is necessary in order to imply the further maximum likelihood estimation of the rural households' technical efficiency. The parameter γ is limited between 0 and 1, and is useful to search for an appropriate starting value for the implementation of an iterative maximization procedure of the maximum likelihood procedure of efficiency estimation, which will be explained further later on. The log-likelihood model's function is described in detail in the attachment to Battese and Coelli (1992, p159-163).

The three-step estimation method, implemented by the FRONTIER 4.1., is described as follows:

1 The Ordinary Least Squares (OLS) estimates of the Cobb-Douglas production function were obtained during the *first stage*. Estimators of β were unbiased, with the exception of the intercept values.

2 The *second procedure* was the grid search of the γ values. β parameters values (except β_0) were set from OLS estimations, while β_0 and σ^2 were defined through the adjustment of the corrected OLS (or COLS) procedure (Coelli, 1995, p247-268). A grid search was implemented in two phases to reach a higher accuracy level of the starting point. As a first step, which was specified earlier, a search for the γ parameter was conducted within the limits of 0.1 to 0.9. The search step was fixed with an accuracy of 0.1. The next phase was a grid search for the starting values obtained from the first phase. As a result of the second phase, a starting value for γ was received with an accuracy of two decimal points.

3 The *third stage* consisted of the ML estimation of the γ values. Many methods of calculating maximum-likelihood estimates need second partial derivatives. The Quasi-Newton method was selected because it only demands the vector of the first partial derivatives. Of the Quasi-Newton methods, the Davidon-Fletcher-Powell method was chosen, as it is widely used by econometricians and also empirically applied by Pitt and Lee (1981) for the analysis of the Indonesian weaving industry's technical inefficiency. The reasons for the selection of the calculation method and of the part of the program code, which implements the iterative procedure calculations (in FORTRAN computer language), was derived from Himmelblau (1972) (Coelli, 1996b, p13). The iteration procedure used the γ starting values defined previously in the grid search. The software

then updated the vector of γ values using the Davidon-Fletcher-Powell method. The procedure terminated at the point of reaching one of the following conditions:

[-] the convergence criterion was implemented (if change between the likelihood function and the parameter was less than 0.00001),

[-] the maximum number of iteration was limited to 100.

The same production variables set, as was defined earlier, was implemented here. The results of the maximum likelihood estimations of the Cobb-Douglas stochastic frontier model are presented in Table 6.6. Some implications might be useful to us. The coefficients' sums of the labor and capital explanatory variables are quite close to 1. This means that the estimation of the main production factors is not far from the proposed theoretical value. However, contrary to the traditional results, the capital factor influence prevails over that of the labor force. Another important result is the high value of the γ estimation (0.82), which means that the main source of the variation in the composite error term is caused by inefficiency. However, significant standard error values certainly deteriorate the level of accuracy of these statements.

Table 6.6: ML Estimates of the CD Inefficiency Stochastic Frontier Production Function for Rural Households in Talas Oblast

Variable	Parameter	Cobb-Douglas SFA Model*
Constant	β_0	0.17
		(0.25)
log (Labor)	β_1	0.15
		(0.79)
log (Capital)	β_2	0.88
		(0.43)
	σ^2	0.74
		(0.95)
	γ	0.82
		(0.53)
Log (likelihood)		-0.25
	N	289

Source: ML Estimations Result from Survey Data
*Estimated standard errors are given in parentheses.

The distribution and descriptive statistics of the rural households' efficiency estimations are presented in Figure 6.8. Variations in the households' efficiency level are significant - from 12% to 90% - and caused by the fact that all rural population is sampled by the survey.

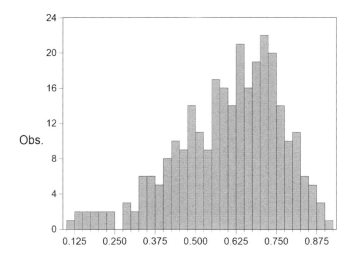

Figure 6.8: Distribution of the SFA Rural Households Efficiency Estimates
Source: Estimations from Survey Data

Unlike traditional estimations of the efficiency of a certain type of firm, sectors or specialized producers, the population demonstrates a wider spectrum of activities besides agricultural production as well as different scales of production. The average efficiency of the rural households is equal to 60%. The level of production efficiency of the same sample with the SFA is significantly higher than the estimates received through the DEA application (See Figure 6.4 on p. 110). We suppose that the distributional assumptions of the SFA specification's error term mentioned earlier, as well as the deterministic requirement of the DEA approach, affect the received results.

The level of production efficiency of the same sample with the SFA is significantly higher then estimates received through DEA application (See Figure 6.4 on p. 110). We suppose that the mentioned earlier distributional assumptions of error term of SFA specification, as well as the deterministic requirement of DEA approach, are affected the received results.

The left-side skewness of the slope of the "low" efficient producers demonstrates that a significant number of households cannot increase efficiency or compensate for it by other income sources. At the same time, the group of "high" efficient producers tensely competes for a productivity increase. The potential for an increase of technical efficiency is high, however, constraints of the households with regard to its increase need to be clarified additionally. Due to the specific character of the rural producers' motivation and capacity to improve, efficiency might be different. The obtained coefficients of the stochastic parametric estimations of the technical efficiency, together with the results of the non-parametric deterministic approach are used further as explanatory variables (determinants) of rural poverty on the household level.

6.3 Poverty - Efficiency Linkage Model Results and Discussions

The last section of this chapter starts with the adoption of the poverty indicator appropriate for use at the household level. The proposed poverty index is estimated with a simple and multivariate cross-sectional model including production efficiency coefficients. Additionally, a possible endogeneity of the technical efficiency estimates with the dependant variable (the poverty index) is also tested. A discussion and interpretation of the results ends the section.

6.3.1 Poverty Indicator Adjustment

The initial problem in modeling the poverty - efficiency linkage is the absence of the appropriate poverty indicator oriented directly to the household level. Generally, poverty analysis aims at defining the poverty indicators on the country or regional level, and the analysis of the poverty incidence and depth of the poor strata of the population. Thus, the necessity of adjusting the poverty indicators for our research purposes was obvious.

With regard to the adoption of poverty indicators, we need to have the opportunity to apply them to all the samples and not only to the poor population. We propose to use the transformed Poverty Gap Index (PGI) indicator for our purposes (see 3.2. on page 20). In its original form, it is applied to those households, whose consumption falls below the poverty line, and to measure the mean income of the poor strata of the population. For our purposes, we use an alternative PGI, which brings us the same results as the PGI 3.2., but with easier calculations, which are given in the Poverty Manual (Haughton, 2005, p72):

$$\mathbf{P}_1 = \frac{1}{n} \sum_{i=1}^{n} \frac{z - y_i}{z}, \text{ if } y_i < z \qquad (6.11.)$$

where y_i is the income of the i^{the} household, z is the poverty line, and n is the number of households in the sample. Only income of the poor households is analyzed in this formulation, for all non-poor households the PGI is considered to be equal to zero. We suggest to introduce the transformation of this formula in order to not calculate the gap, but the

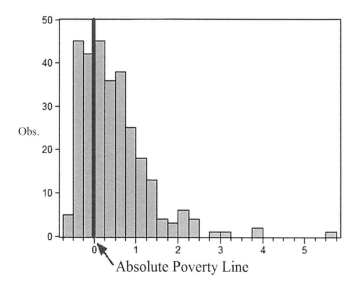

Absolute Poverty Line

Figure 6.9: Distribution of the PDI for the Talas Oblast's Sample
Source: Estimations from Survey Data

distance of each household to a poverty line as follows:

$$\mathbf{P}_{1i} = \frac{y_i - z}{z}. \tag{6.12.}$$

Changing the position of the income and poverty line in the equation affect the character of an index. An index, transformed in this way, measures the distance of the income (in our case the per capita consumption) of the i^{the} household from the poverty line. If the income is higher than the poverty line, it will demonstrate a non-negative value and if it is lower than the poverty line, the value of the index will be negative. Thus, the well-being of all the sampled households may be analyzed relative to a poverty line, either below or above it.

The suggested name for this household based index is Poverty Distance Index (PDI). The PDI may be applied to absolute and extreme poverty lines; however, we only use it for the absolute poverty line, because further implemented regressions show very similar results for both of the indexes. Descriptive statistics and distribution of the Poverty Distance Index for the Absolute Poverty Line (PDI) are given in Fig. 6.9. The sample's main share is located in the zone close to the poverty line, with a certain shift of the mean above zero values. Vulnerability of the population to external price volatility, inflation and exchange rate stability is clear from the character of the well-being's distribution among the rural strata of

119

the population.

6.3.2 Poverty - Efficiency Model Results

The initial stage of modeling consisted of an explanatory linkage of the efficiency estimates with the proposed poverty index. Modeling started with a quite primitive linear regression model:

$$PDI = \beta_0 + \beta_1 \cdot TE + \varepsilon, \tag{6.13.}$$

where the dependant variable PDI is a Poverty Distance Index and TE is an efficiency estimate, obtained from DEA and SFA efficiency measurement, described in sections 6.2.2 and 6.2.3. The aim of this exercise was to test all of the proposed efficiency estimators as poverty determinants. Eight different estimates of simple regression analysis results are given in Table 6.7.

The results of SFA and DEA estimations need to be analyzed separately. The SFA inefficiency estimator demonstrates a high level of the relationship between poverty and efficiency on the household level. The increase of the efficiency level by 1 percentage point improves the PDI to 1.1 percentage point. However, the explanatory power of the model is quite modest (5% only). The necessity of applying the multiple regression model is conclusive.

The DEA model opens some deeper level of relationship due to the possible breaking up of the aggregated efficiency measure into its parts. DEA at constant returns to scale (column 2) defines the bigger predicted value of the dependent poverty variable growth from efficiency increases, i.e. 1.3 percentage points, in the case of stability of other factors. The residual sum is smaller than the SFA model, due to a higher explanatory power (14%), which is better than the SFA relationship, but still insufficient in itself.

The next three columns define the importance of capital efficiency at variable returns to scale on the households' poverty level. The results of this group of estimators' correlation repeat the conclusions made before in the analysis of the capital scale efficiency's distribution (see Fig 6.5 on page 111). The DEA scale efficiency estimator, measuring the capital efficiency (column 5), demonstrates a negative influence on poverty. It means that the current scale efficiency, in the case of capital input, is so effective that an increase of the capital input scale efficiency will lead to the subsequent decrease of the PDI. This indicates that poor households operate at a more effective level than non-poor ones and that the limits of the increase of capital efficiency for the majority of the households are hardly applicable for such a micro-scale level of production. The explanatory power of the DEA capital based model is insignificant.

Table 6.7: Preliminary Regressions of PDI to Efficiency Estimations

Dep. Var.	PDI^*							
No. obs.	289							
Const.	-0.24	-0.03	0.30	0.18	0.93	0.19	0.10	0.05
	(0.2)	(0.1)	(0.1)	(0.1)	(0.2)	(0.1)	(0.1)	(0.1)
Stochastic Frontier Estimates								
SFA	1.15	-	-	-	-	-	-	-
	(0.28)							
Input-Oriented DEA Frontier Estimates								
$DEA1_{CRS}$	-	1.33	-	-	-	-	-	-
		(0.20)						
DEA Capital Scale Efficiency								
$DEA2_{CRS}$	-	-	0.73	-	-	-	-	-
			(0.36)					
$DEA2_{VRS}$	-	-	-	0.91	-	-	-	-
				(0.22)				
$DEA2_{SCCAP}$	-	-	-	-	-0.58	-	-	-
					(0.19)			
DEA Labor Scale Efficiency								
$DEA2_{CRS}$	-	-	-	-	-	2.95	-	-
						(0.46)		
$DEA2_{VRS}$	-	-	-	-	-	-	0.89	-
							(0.22)	
$DEA2_{SCLAB}$	-	-	-	-	-	-	-	1.73
								(0.25)
R^2	0.05	0.14	0.01	0.05	0.03	0.13	0.05	0.14

Source: OLS Estimations Results from Survey Data
*Notes:
a) Standard errors are given in parentheses.
b) All the given coefficents are significant at least at a 5% level.

The last three columns explore the next source of potential improvement of the efficiency of rural households, the labor force. The best prediction power shows an improvement of the labor force's efficiency at CRS: each percentage point of increase of the efficiency of using labor could bring us an increase of the poverty index of approximately 3 percent. The explanatory power of the model is better than the capital input scale efficiency, but really close to the first DEA model at CRS. We may conclude that the important result of DEA estimation is the clarification of the labor-capital interrelation role as a main potential source

121

of the impact on the poverty indexes through an increase of production efficiency.

The main conclusions of the preliminary relationship analysis of the poverty and efficiency estimations are the following:

- the following efficiency estimations are appropriate for the multivariate poverty determinants analysis - SFA, $DEA1_{CRS}$, $DEA2_{SCLAB}$,

- the explanatory power of the models is low and the confidence intervals of efficiency estimates are large,

- the influence of labor and capital factors is important in order to understand the poverty and efficiency relationship.

An analysis of the results of the initial modeling of the poverty-efficiency linkage brings us to the following logical step, i.e. the multivariate regression analysis. The following, additional factors, which improve the prediction power of the poverty estimation besides efficiency indicators, were included in the model:

1 Demographic variable - The Household Size variable includes the amount of all members of the household ($HH\ SIZE$), measured in persons.

2 Human capital investment variables:

[a] Medical Expenses take into account the annual sum of expenditures of all medicines and medical treatments received and services employed by household members ($MED\ EXP$), measured in thousand som per person.

[b] Education Expenses include all expenses of the household on education in primary, secondary and higher education institutions within the last academic year ($ED\ EXP$), measured in thousand som per person.

3 Capital variables:

[a] Savings indicate the savings of a certain household (SAV) in form of cash, deposits or borrowing to other people, measured in thousand som per person.

[b] Land Used by Household takes into account all the land resources used by the household for the purpose of agricultural production (includes irrigated and non-irrigated main land plot, kitchen garden, hayfields, gardens, tenured land plots, with exclusion of land rented out to other people or given to them for free) ($LAND\ USED$), measured in hectares per person.

4 Behavioral variables:

[a] Celebrations include all expenses of the household made in form of gifts, costs of celebrations, funerals, weddings and all types of celebrations in rural society within the last year ($CELEBR$), measured in thousand som per person.

[b] Support to Relatives includes all expenses paid for children, parents and other close relatives in the form of money, products and other goods for the last 12 months ($RELATIVES$), measured in thousand som per person.

The final specification of the model is presented as follows:

$$PDI = \beta_0 + \beta_1 \cdot TE + \beta_2 \cdot HH\ SIZE + \beta_3 \cdot MED\ EXP +$$
$$+ \beta4.ED\ EXP + \beta_5 \cdot SAV + \beta_6 \cdot LAND\ USED + \qquad (6.14.)$$
$$+ \beta_7 \cdot CELEBR + \beta_8 \cdot RELATIVES + \varepsilon,$$

where TE - denotes the three estimates of the technical efficiency for rural household SFA, $DEA1_{CRS}$, $DEA2_{SCLAB}$, and all other explanatory variables defined earlier. Thus we calculate three variants of the model for different frontier estimates - one for the SFA approach and two for the DEA specification.

The results of modeling are presented in Table 6.8 on page 124. The inclusion of the additional explanatory variables expectedly decreases the linkage power between the explanatory relationship of the efficiency estimates and poverty index. For technical efficiency estimates SFA and $DEA1_{CRS}$, the level of predicting power decreases significantly, a 1 percentage point increase of efficiency could increase the PDI to 0.6 and 0.4 percentage points, respectively. The labor scale efficiency estimate, $DEA2_{SCLAB}$, decreased from 1.7 percentage point to 1.1 percent only. Confidence intervals of efficiency estimations are kept significant - t-statistics values of efficiency estimates vary from 4.3 to 6.

Among the proposed explanatory variables, the most significant power is demonstrated by land which the household actually uses for agricultural production. An increase of the size of this land per capita on one percentage point will lead to a growth of the PDI by 0.18 percentage points in the case of the DEA Labor Scale model, by 0.37 percentage points for the DEA CRS model, and by 0.47 percentage points for the SFA. The influence of this parameter may indicate information pertaining to the relationship between the main limited production factor and the poverty index, which is not covered by the efficiency of agricultural production. It is interesting that livestock, which is included in the model, directly affects the efficiency estimates confidence intervals. The DEA estimator of the labor factor's scale efficiency (column 3) has the weakest effect of land factor influence. Its statistical influence is significantly lower, and confidence intervals are weaker, than in the case of DEA at CRS and SFA specifications, which includes land in their capital factor.

Size of the household is the only negative factor in the model in relation to the poverty index.

Table 6.8: Multivariate Regressions of Poverty Distance Index

Dep. Var.	Coef.	PDI*		
No. obs.		289		
Const.	C(1)	-0.09	0.15	0.34
		(0.12)	(0.09)	(0.08)
Efficiency estimates				
SFA	C(2)	0.62	-	-
		(0.145)		
$DEA1_{CRS}$	C(2)	-	0.49	-
			(0.11)	
$DEA2_{SCLAB}$	C(2)	-	-	1.10
				(0.18)
Demographic variables				
$HH\ SIZE$	C(3)	-0.09	-0.09	-0.12
		(0.01)	(0.01)	(0.01)
Human capital investment variables				
$MED\ EXP$	C(4)	0.10	0.09	0.10
		(0.01)	(0.01)	(0.01)
$ED\ EXP$	C(5)	0.06	0.07	0.06
		(0.01)	(0.01)	(0.01)
Capital variables				
SAV	C(6)	0.05	0.04	0.04
		(0.001)	(0.01)	(0.01)
$LAND\ USED$	C(7)	0.47	0.37	0.18
		(0.06)	(0.06)	(0.08)
Behavioral variables				
$CELEBR$	C(8)	0.08	0.09	0.08
		(0.00)	(0.00)	(0.00)
$RELATIVES$	C(9)	0.04	0.05	0.04
		(0.00)	(0.00)	(0.00)
R^2		0.774	0.775	0.787

Source: OLS Estimations Results from Survey Data
*Notes:
a) Standard errors are given in parentheses.
b) All the given coeffcents are significant at a 1% level, except for the $LAND\ USED$ specification in the third column, which is significant at a 5% level.

The decrease of the household size by 1 percentage point may bring about 0.09 to 0.12 percentage points of PDI growth. Statistically, the influence is not that highly significant, but consistent in all three models. Other explanatory variables demonstrate quite a modest positive influence, and vary from 0.04 to 0.1 percent of the positive influence on the poverty index for each percentage point of each of the appropriate variables - $MED\ EXP$, $ED\ EXP$, SAV, $CELEBR$, $RELATIVES$, if all other conditions remain constant.

The influence of the increase of medical expenses and expenses on education on poverty is not obvious. On the one hand, the correlation between the PDI and those expenses may be explained by the importance of the investment in health and knowledge. It may have led to an increase of the working and mental capacities and additionally support the increase of efficiency. Consequently it may have improved the poverty index. However, it may also have signalled that more educated and healthier households with a higher welfare level tend to spend more on medical and educational purposes. It means that the linkage in this case has a more demonstrative character and that the increase of a household's welfare may tend to increase these expenses. We can apply the same conclusion to the correlation linkage between PDI and behavioral variables (expenses on celebrations and support to relatives). Wealthier households may tend to spend more on those expenses, then poor ones.

Unlike the mentioned variables, savings formally stay out of the consumption aggregate and thus cannot be directly linked with the PDI. Nevertheless, we can expect that there is a higher probability for the presence of savings and the size of savings for non-poor households than for the poor ones. Thus, the causality of the variables' relationship again is not clear.

Besides the efficiency estimates and land factor used by the households variable, other factors play a role of supplementary forces in the model. Nevertheless, the inclusion of those variables significantly support an increase of the explanatory power of the model. The sizes of the R^2 vary from 77% to 79% and generally explain the relationship of poverty with the introduced explanatory variables set. The model's results support the original idea of an existing relationship between agricultural production efficiency and poverty level.

Another important implication we may derive from the model results is the definition of the source of potential efficiency: the labor factor. Scale efficiency of the labor force at variable returns to scale shows a better performance than other efficiency estimations including the capital factor in the production efficiency estimation. Moreover, the linkage of the efficiency with poverty does not contradict to the possible explanation that an investment in health and education may lead to a higher production efficiency of labor.

Despite the good results, which the model shows us, we additionally need to stress the poverty-efficiency linkage. In the proposed specification, the results of the model application may be deteriorated by the possible endogeneity of the technical efficiency variable. The risk that the omitted variable influences the efficiency index is possible due to the origin of the efficiency estimates. As we already know from the theoretical background of the frontier method, the deterministic specification in the DEA approach supposes that there are no measurement error. This means that DEA efficiency estimates may include a measurement error term. For the SFA method we need to stress the importance of the distributional assumptions of the error term, which certainly affects the solidity of the results. Thus, the possibility of a confounding 'unknown' parameter's existence, which correlates with the error term, efficiency estimates, and separately affects the explanatory parameter (poverty index) is plausible.

One of the obvious implications of the endogeneity of the poverty index linkage with the efficiency level may be a potential 'poverty trap' of the poor households. Besides the size of the capital factor, poor households may presumably be limited in their attempts to increase productivity, because of the absence or insufficient access to the productivity-increasing inputs - high-quality seeds, fertilizers, machinery services, etc. Thus, wealthier households have a competitive advantage and can be more efficient than the poorer ones simply because of their better access to inputs. In such a situation, poor households tend to be less efficient producers, irrespective of their efforts in the production.

In order to solve the problem of endogeneity of the efficiency variable, we decided to introduce a model with an instrumental variables regression - the two-stage least squares (TSLS) method. We suppose that there are variables, which are exogenous from the dependant variable - the poverty index and the error term, but may affect the technical efficiency level. From the multivariate regression we learned that the poverty index is sensitive to the purely capital factor - the size of the used agricultural land per household member. We suggest to use qualitative parameters for the land, irrespective of the wealth index of the households. On other hand, land characteristics may directly affect the production performance of the household, and the free of error term assumptions of the frontier model.

Additionally it has to be noted that the land reform in Kyrgyzstan was implemented quite recently, at the end of 1990s. Land privatization was performed in the form of an equal distribution of the available agricultural lands among the rural households in the respective rural areas. The land plot's size was defined on a per household basis, and depended on the size of the household. The land share was defined separately for each rural area, as the product of the division of 75% of the available agricultural land by the number of all rural residents, including children, registered in the respective area. Thus big families received bigger land plots, in proportion to the household size. Researchers have noted that the land distribution implementation was quite fair (and transparent). The possible land resale by the poor people was restricted by the imposition of limitations on the land market. The moratorium on land sale was terminated in 2001, but other rules limits the land redistribution is valid up to now. Thus land was distributed more or less randomly and in proportion to the size of the household among the rural population (Lehrman and Sedik, 2009, p3-4,14-15).

A number of additional qualitative variables for the land factor were introduced to the research questionnaire, to be applicable as potential *instrumental variables* (IV). Two of the factors, which were selected finally as IV, were land quality of the main land plot and distance to the main land plot from the house. Thus, we suppose that the proposed instruments explain the efficiency estimates and do not depend on the poverty index, as well as on error assumptions. The TSLS estimation method was applied, with the mentioned instruments, as follows:

1 Land quality - this variable assessed the quality of the main field land plot (scale from 0 to 4):

[a] 0- do not know / or not applicable if there are no land plot,

[b] 1- bad quality,

[c] 2 - sufficient (average) quality,

[d] 3 - good quality,

[e] 4 - very good (excellent) quality.

2 Distance to the land plot - variable defining the distance between the place of residence and the field/land plot(scale from 0 to 4):

[a] 0 - do not know / or not applicable if there are no land plot,

[b] 1 - less than 1 km,

[c] 2 - from 1 to 5 km,

[d] 3 - from 5 to 10 km,

[e] 4 - more than 10 km.

For each of the options we constructed a categorial dummy variable, which was equal to 1 if the condition was fulfilled and 0 if it was not. The data was available for all households, because it also included the households which did not have their own land or do not use it. Results of the TSLS are presented in Table 6.9 on page 128. The introduction of the IV affected the results of the model. The technical efficiency's statistical significance increased in comparison with the OLS specification (see Table 6.7 on p. 121). At the same time, the error term increased and, consequently, the confidence intervals decreased. This, however, did not gave give us a reason to state that the results were significant from a statistical point of view. The explanatory power of the model differed. For the model with SFA efficiency estimator, the explanatory power of the model became negative, and we did not report it.

Table 6.9: TSLS Estimation Regressions of PDI and Efficiency Estimations

Dep. Var.	Coef.	PDI*		
No. obs.		289		
Const.	C(1)	-1.17	-0.23	0.04
		(0.53)	(0.19)	(0.14)
Efficiency estimates				
SFA	C(2)	2.68	-	-
		(0.88)		
$DEA1_{CRS}$	C(2)	-	1.88	-
			(0.53)	
$DEA2_{SCLAB}$	C(2)	-	-	1.76
				(0.56)
R^2		–	0.11	0.14
Instrumental variables:				
LAND QUALITY		DISTANCE LAND		

Source: TSLS Estimations Results from Survey Data
*Notes:
a) standard errors are given in parentheses.
c) the results of the first-stage LS estimation are presented in the Appendix E.

The DEA specification at the CRS decreased R^2 by some percent, while the DEA labor scale efficiency estimator's explanatory power remained on the same level.

The TSLS model shows that the inclusion of the instruments still supports the original idea of the linkage of the poverty index with the production performance of the household. The results, however, are not equally valid for DEA and SFA estimates. DEA estimators, especially the labor scale efficiency specification show higher results than other specifications. Such a behavioral relationship may be explained by the advantages of the DEA non-parametric deterministic approach, as well as the possibility to exclude the influence of the capital factor from the model, whose exploitation is limited at the current production scale. The SFA estimator shows weaker results, which again might be explained by the limitation of the model: distributional assumptions of the error component seriously increase the risk of endogenously confounding the efficiency parameter, primarily with the error term. The loss of explanatory power of TSLS with the SFA estimator may also be caused by the increase of the error term of the proposed IV. We may argue that a farmer's assessment of land quality is subjective, as well as the assessment of the distance to the land, which is based on an approximation and cannot be absolutely true. Therefore, the potential error term may cause a weakening of the SFA estimator behavior.

Finalizing the discussion, we may conclude that at least some of the proposed frontier specifications support the validation of the relationship between poverty and efficiency on the household level. Results of the statistical estimations show that efficiency improvement will bring a consequent increase of the welfare of the rural households, as measured by the proposed Poverty Distance Index. The final conclusion and proposed policy recommendations are given in the last part of the work.

7. Conclusions

At the beginning of the research, the main idea was to define the driving factor, which led to the rapid decrease of poverty in the period from 2001 to 2007. Defining such a concept meant to understand the core problem, i.e. the constraint of the potential poverty decrease observed since 2008, especially in the rural areas. The hypothesis of a poverty-production efficiency linkage for the rural population was defined as the answer to this problem after an analysis of the post-communist transformation reform of the rural sector in the country. As a result of the land privatization, almost the whole rural population became individual rural producers. Concerning this matter, a similar explanation of the rural poverty determinant on a household level was not found in the literature.

The second result was that the current study produced a unique primary data set, collected within the field survey. The survey methodology was arranged to provide a high quality of the data from the selected sample, and some specific information (e.g. land quality parameters data) is also unique, at least in the case of Kyrgyzstan. The collected data set consists of comprehensive information, which allows us to implement such different tasks as poverty measurement, and a production efficiency estimation on the household level.

Empirical results demonstrate the existing and statistical significance of the poverty-efficiency linkage in the case of simple and multivariate linear regression. The possible endogeneity of this linkage is resolved with the proposed instrumental variables, not correlating with the wealth of the household, but affecting the technical efficiency. Modeling of the production technology's limits also provides us with additional information about an overuse of the capital factor and potential productivity sources.

However, we need to define a number of problems we faced in our work, but cannot solve due to limits of time and available resources, and therefore need to address in future studies. The SFA approach shows that confounding the error term, omitted in the efficiency estimator, makes the SFA approach an insufficient method for the validation of proposed idea with the available cross-section data. Nevertheless, SFA development suggests potentially more solid results gained from the panel data, the same can be said for the DEA approach. Thus, our analysis only constitutes the first steps in the future exploration of this issue.

Besides the data availability, our analysis is also limited by the selected target region, and it is obvious that it is necessary to expand the research to the national level to validate the poverty-efficiency linkage for a whole country. A linkage is obviously play an important role in those regions where non-farm income activity is limited, while available capital factors (agricultural land, pastures and water) are still available for the most of the population. For the regions, where production resources are less available, other regional peculiarities may be more important in the poverty alleviation context.

We also may state that the proposed instrumental variables used in TSLS are not perfect in themselves due to a potential measurement error. Moreover, we did not specify good instrumental variables for the labor force. Thus, the issue of the search for an optimal variable set is an open issue, too.

The proposed idea reflects the unique situation of the agricultural production in Kyrgyzstan, but we cannot be sure that the proposed linkage will be stable in the future. There are a number of factors, which may affect this process and such linkage may be terminated with the change of the agricultural production's character. The issue also needs to be explored in the direction of the idea of a potential application in other developing countries where such a micro-scale household-based production exists.

An implication of the proposed idea is the potential number of ways by which we could increase productivity and thereby improve the welfare of the rural population. We need to specify that the improvement will affect the whole population as we have modeled the situation for the full sample, and not only for the 'poor' population strata. In this regard we need to point out that, due to the proposed PDI, the distribution of the consumption concentrates closely around the poverty line (see Figure 6.9 on page 119). Thus we may not simply support the decrease of the poverty situation for the whole rural population, but also may potentially decrease the vulnerability of those households which are located closely above the poverty line. To specifically address the problem of the support of the poor strata within the context of the proposed idea, we need to specify additionally that targeting poverty alleviation is a specific part of the public policy recommendation.

Recommendations for the national government are only general directions concerning the questions where an improvement of production efficiency with the purpose of poverty allevia-tion could be effectively implemented. One of the first obvious directions we need to address is the issue of restructuring the analytical framework for the poverty fighting strategy. It is obvious that the living standard information, collected by the NSC, was not fully explored. Potentially, KIHS may be a more effective instrument for policy recommendations, coupled with an inclusion in a process of data analysis performed by academic institutions and uni-versities. The creation of a more transparent atmosphere, the exchange of information for free, and open discussion may bring much more advantages for the government.

National ministries and agencies, responsible for the rural sector, are obviously faced with the absence of finances to solve the productivity problem, but also with insufficient state regulations in this sector. Rural households (farmers) are independent from the state in the issues of production and technologies. Thus, only the motivation to use modern technologies may bring positive results. Nevertheless, it needs to be stressed that state selection farms need to suggest a type of new seeds (e.g. haricot beans) for Talas Oblast, which have been adapted to the local conditions, and to disseminate the results through a number of demonstration sites, which is not the case at the moment. Personal experience shows that rural inhabitants actively use a copy strategy, applying what they think is the best practice. Therefore it is important to cooperate with the best farmers in each district to disseminate the results widely among the farmers. The same is true for the use of fertilizers, plant protection systems or simple agrotechnical methods.

There is a serious constraint with regard to the realization of an economy of scale in agricultural production in Kyrgyzstan, which seriously impairs the effective use of modern technologies. The necessity to introduce cooperation among farmers is obvious. However the memory of the inefficiency of collective work in Soviet times is still strong among farmers and agricultural cooperatives are quite rare in the country. Nevertheless, this is still the area in which the Government can create more attractive conditions for agricultural cooperatives, including schemes of leasing agricultural machines and supplying fertilizers, seeds and other inputs. However, the exact strategy of how to do so needs to be studied separately.

Special attention needs to be devoted to the support of the poor population. Poor households are limited in the potential improvement of their production efficiency. Currently, state support is basically limited to the free distribution of food products or some subsidy for poor households. But those households might be supported more effectively by a wider access to the more productive inputs. Support may be provided in those cases in which households can afford a production activity, and should be complex, i.e., covering all the aspects of productivity improvement. This assistance should in turn be supported at the local community level. The detailed schemes of such a support also require a special study with tests in pilot villages to develop optimal ways of such a support.

As the final comment, it is necessary to repeat that the main task of the present study was performed. Some issues, as method limitation and data availability, which have arisen from this work will be addressed in future studies, also by the author. Despite the fact that the findings are mainly practical and targeted at state agencies and anti-poverty institutions, they may also be useful for other researchers, working in the similar field.

A Research Questionnaire

This Appendix presents the research questionnaire used during the full field study in Talas Oblast. Two versions of the document are included the Kyrgyz version (main field document) as well as the English version.

Kyrgyz Version

Талас Областы   *Маалыматты чогултуучу сырды сактоого кепилдик берет*

ҮЙ ЧАРБАСЫН ТАНДАП ИЗИЛДӨӨ

ҮЙ ЧАРБАСЫ
ЖАЙГАШКАН ЖЕР_____(ЭЛ ЖАШАГАН ПУНКТ, КӨЧӨ. ҮЙ)

АЙЫЛ ЧАРБАСЫНЫН КОДУ

Интервьюердин коду

Натыйжанын коду

10-интервью бүттү
11-интервью бүткөн жок
20-квартира (үйүндө) жок
21-турак жай бузулган

22-коммерциялык
Мекемеге айландырылган
30-жашагандар менен контакк
түзө албадым
31-жооп берүүдөн баш тартты
40-жашаган жерин кторгон
41-башка

Интервью өткөргөн
датасы   **күнү |__|__| айы |__|__| 2011ж.**

ИНТЕРВЬЮ өткөрүү убактысы

Башталышы: |__|__| саат |__|__| минут

Аягы: |__|__| саат |__|__| минут

- 1 -

136

Маалыматты чогултуучу сырды сактоого кепилдик берет

Үй чарбасын тандап изилдөө

I бөлүм - **Үй чарбасынын карточкасы**

Үй чарбасы жайгашкан жер________________________________
(калктуу пункттун аталышы)

Үй чарбасынын коду ☐☐☐☐☐☐

Интервьюердин коду ☐☐☐☐

Үй чарбасынын контролдук карточкасы

Форма №1

№ п/п	1. Аты жөнү *Интервьюер, суроолор 2-13 өтөөрдөн мурда үй чарбасында жашагандардын баарынын тизмесин алыңыз*	2. Жынысы эркек.- 1 аял - 2	3. Айыл чарбасынын башчысына карата мамилеси Башчысы ……………1 Жубайы……………2 Уулу /кызы……………3 Күйөө баласы /келини……4 Атасы/апасы ……………5 Родители супруга (и)……6 Агасы /эжеси……………7 Небереси……………8 Чоң атасы/чоң энеси……9 Жээндери……………10 Иниси /карындашы Аялы /күйөөсү……………11 Башка туугандары ……12 Малай……………13 Ижарачы……………14 Башкалар……………15	4. (Аты)Сиздин үй чарбасынын мүчөсү болуп саналабы? Ооба……………1 Жок……………2	5.Туулган күнү Маселен 01.05.1995			6. Толук жашы	7. Үй бүлөлүк абалы Катталган никеде турат……………1 Катталбаган никеде турат……………2 Ажырашкан……………3 Бөлөк турат,бирок Ажырашкан эмес……4 Жесир……………5 Никеге эч качан Турган эмес……………6 13 жашка чейин……………98	8. Улуту Кыргыздар……1 Орустар……2 Украиндер……3 Өзбектер……4 Казактар……5 Белорустар……6 Тажиктер……7 Татарлар……8 Дунгандар……9 Башкалар……10
					күнү	айы	жылы			
1										
2										
3										
4										
5										
6										
7										
8										
9										
10										
11										
12										

Талас Областы

Маалыматты чогултуучу сырды сактоого милдеттуу берет

№ п/п	1. Аты жону	10. Иш менен алектенгендердин статусу Жалданып иштегендер.................1 Индивидуалдуу негизде иштегендер...........2 Фермер.....................2.1 Иштетен ишкананлар.......3 Жалпы бюджеча ишчи.........4 Майыгыгы боюнча пенсия........5 Багуучу адамын жоготкондугуна байланыштуу пенсия............6 Эмгектин отунн шарты пенсия...........7 Боюнча пенсия........8 Аскер кызматкынлыгынра боюнча пенсия........8 Студенттер.......9 Иштетен студенттер......10 Окуучулар...........11=>к.12 Жумушсуздар..........12 Иштетен окуучулар..........13=>к.12 Эч кандай статус жок....................99=>к.12	11. Жекече эмгек ишмердуулугунун киреши же эмгек акысынын, пенсиянын, стипендиянын өлчөмү канда? (сом) Эмгек акысы..................1 Жекече эмгек ишмердуулугунун тушкон киреши.......2 Пенсия ....3 Стипендия 4	12. Акыркы уч айда (Ата) уй чарбасынла туруктуу жашаган Ооба.....................1 интернынуын анты Жок.....................2	13. Акыркы уч айда(Ата) Канча куп жок болду? Кундун эсеби
			Сумма (сом)	**Код**	
			Сумма (сом)	**Код**	
1					
2					
3					
4					
5					
6					
7					
8					
9					
10					
11					
12					

ИНТЕРВЬЮЕР, ЧАРБАСЫНДА ЖАШАГАНДАРДЫН ЖАЛПЫ САНЫН ЖАЗЫҢЫЗ _____
ФОРМДОН ИЧИНЕН:
    УЙ ЧАРБАСЫНДА БОЛГОН МУЧОЛОРДУН САНЫ _____
    УЙ ЧАРБАСЫНЫН УБАКТЫЛУУ ЖОК БОЛГОН МУЧОЛОРУНУН САНЫ _____
    БӨТӨН АДАМДАРДЫН САНЫ _____

138

Талас Областы

Маалыматты чогултуучу сырды сактоого жеңилдик берет

II БӨЛҮМ - ҮЙ ЧАРБАСЫНЫН МҮЧӨЛӨРҮНҮН БИЛИМИ

[ИНТЕРВЬЮЕР] Аты жөнүлгу тизмесин контролдук карточка жана номерлерин тууга келишин текшириңиз

Контролдук карточканын катар N	Аты	1. Сиздин билимиңиз кандай?	2. Кандай кайсы чөйрөсүн изилдегенсиз(изилдеп ет жатасыз)?	3. Азыркы учурда сиз окуйсузбу?	4. Сиз кайсы окуу жайында окуйсуз ?	5 Жалпы билим берүү мекемесинин тиби	6 Канчанчы класста окуйсуз?	7. Кандай себеп менен окубайсыз?
1								
2								
3								
4								
5								
6								
7								
8								
9								
10								
11								
12								

- 4 -

139

Талас Областы

Миграцияны чагылтуучу сырды сактоого кепилдик берет

III БӨЛҮМ - МИГРАЦИЯ

№ п/п из контролдук карточкасына	1. Сиз кайсы өлкөнүн элкопту жаранысыз?	2. Сиз жашаган жериңиз де тородуу жуз беле?	3. Сиз кайсы өлкөдө төрөлгөнсуз?	4 Картага тандаган кайсы областын да тородгон жериңиз сүз?	5. Сиз төрөлгөн жер	6. Сиз теред гон жердөн канча жылы нызда көчүп кеттиң низ?	7. Эмне үчүн төролгон жериниздн ташлап кеттиниз?	8. Акыркы 10жылдын ичинде башка жердге уч айдан ашык туруп, ашыр да жашап жаткансызбы?	9. Акыркы жашаган жерге кайдан көчуп келдиниз?	10. Сиз жашаган жер	11. Эмне учун Сиз ашыр жашаган жериниэге келдиниз?	12. Акыркы жоку кочкондон Сиз бул эл жашаган пунктта канчадан бертраажыкй сыз?	13. Сиз жашаган жериниэде каттоду ну1 бе.се?
1													
2													
3													
4													
5													
6													
7													
8													
9													
10													
11													
12													

- 5 -

БИШКЕК 41711 / ЫССЫК-КӨЛ...41702 / ДЖАЛАЛ-АБАД...41703 / НАРЫН...41704 / БАТКЕН...41705 ОШ....41706 / ТАЛАС....41707 / ЧҮЙ....41708

КЫРГЫЗ РЕСПУБЛИКАСЫНЫН ОБЛАСТТАРЫНЫН КОДДОРУ:

140

IV БӨЛҮМ
ТУРАК ЖАЙ ШАРТТАРЫ

1. Сиздин турак жайдын түрү? ♦

Өзүнчө үй...1
Үйдүн бөлүгү.................................2
Башка турак жай..............................3
Убактылуу жай...............................4
Квартира 5
Жашоо үчүн пайдаланылган,
бирок ылайыкташпаган жай 6

2. Сиздин турак жайдын менчик формасы? ♦

Мамлекеттик.........................1
Жеке....................................2
Аренда...............................3⇒ **суроо 4**
Башка................................4

3. Сиз бул турак жайды кантип алдыңыз? ♦

Ордер боюнча алдым..............1⇒ **суроо 7**
Жеке адамдан алдым...............2⇒ **суроо 7**
Курдук...............................3⇒ **суроо 7**
Алмаштырдык.......................4⇒ **суроо 7**
Мураска алдым.....................5⇒ **суроо 7**
Башка.................................6⇒ **суроо 7**

4. Сиз турак жай үчүн аренда төлөйсүзбү? ♦

Ооба...1
Жок..2⇒ 7

5. Арендага ай сайын канча сумма төлөйсүз? ♦

Сумма (сом)...............[   ]
Жооп берүүнү
каалабайм..9

6. Турак жайдын төлөмүнө коммуналдык кызмат кошулабы? ♦

Ооба...1
Жок...2

7. Сиздин үй-бүлө ээлеген жалпы аянт канча?
(1чарчы.м. чейинки тактыкта) ♦

Чарчы метрлер............|_|_|_|,|_|

8. Сиздин үй-бүлө ээлеген турак жай аянты канча? (1чарчы.м. чейинки тактыкта) ♦

Чарчы метрлер ....|_|_|_|,|_|

9. Турак жайыңыз канча бөлмөдөн турат? ♦

Бөлмөлөрдүн саны......................|_|_|

10. Сиз бул үйдө канчадан бери жашайсыз (бул квартирада)? ♦

Толук жылы...........................|_|_|
(эгерде бир жылга чейин болсо-0)

11.Сиздин үйүңүз кайсы жылы тургузулган? ♦

|_|_|_|_| жылы
Жооп берүүнү каалабайт................9

[**ИНТЕРВЬЮЕР!** *12—суроону турак жай 2000-жылга чейин курулган мезгилде гана, берүү керек*]

12. Сиздин үй капиталдык ремонттон өттүбү? ♦

Ооба...1
Жок...2 ⇒ **суроо 14**
Жооп берүүнү
каалабайт...9 ⇒ **суроо 14**

13. Сиз үйүңүздү капиталдык ремонттон акыркы жолу качан өткөргөнсүз? ♦

Капиталдык ремонт
өткөрүлгөн   жыл |_|_|_|_|
Жооп берүүнү
каалабайт 9

14. Сиздин турак жайыңыздын дубал негизги материалы кандай? ♦

Кыш..1
Бетон плитасы..................................2
Бышпаган кыш, саман.....................3
Жыгач, устун.....................................4
Брезент, кийиз...................................5
Топурак, чопо...................................6
Башка (көрсөтүңүз)...........................7

15. Сиздин турак жайыңыздын чатырынын негизги материалы кандай? ♦

Шифер..1
Темир , калай....................................2
Камыш..3
Черепица...4
Жыгач..5
Бышпаган кыш, саман 6
Башкасы(көрсөтүңүз)........................7

16. Шарттардын түзүлүшү: ♦

N	Аталышы	бар-1 жок-2 ⇓ кийин.түрү
1	А	2
1	Жылытуу	
А	Газ	
Б	электр	
В	көмүр	
Г	Башка отун	
2	Телефон	

17. Электроэнергиясына эсентегич орнотулганбы? ♦

Ооба..............................1
Жок................................2⇒ суроо19
Электричество жок ....3⇒ суроо 20

18. Бул эсентегичти Сиздин үй чарба гана колдонобу же башкалар менен бирге колдонобу? ♦

Бир гана үй чарбасы...................1
Башка үй чарбалары менен бирге ...2

19. Сизде акыркы жылы электроэнергиясын канча жолу өчүрлү? ♦

Эч качан өчүрбөйт..........................1
Бир жылда бир нече жолу....................2
Бир айда бир жолу...........................3
Бир жумада бир жолу.......................4
Бир жумада бир нече жолу...................5
Күн сайын...................................6

20. Сиздин үй чарбаңыз суу менен жабдуунун кайсы негизги булагынан пайдаланат? ♦

Суу түтүгү.................................1
Ачык кудук.................................2
Жабык кудук................................3
Жеке менчиктеги колонка..................4
Коомдук колонка............................5
Суу сактагыч, булак, дайра, көл,
көлмө, арык..................................6
Ташылып алынган суу (челек)............7
Башкасы....................................8

21. Суу менен жабдуунун булагы кайда жайгашкан? ♦

Үйдө(квартирада)......................1
Короодо....................................2
Көчөдө......................................3

22. Сиздин турак жай суу менен камсыз кылуу булагынан кандай аралыкта жайгашкан? ♦

100 м. аз....................................1
100-200 м..................................2
200-500 м..................................3
500-1000 м.................................4
1000 м. ашык...............................5

Жооп берүүнү
каалабайт..............................9

23. Акыркы жылы канча жолу муздак сууну өчүрлү? ♦

Эч качан өчүрбөйт..........................1
Бир жылда бир нече жолу....................2
Бир айда бир жолу...........................3
Бир жумада бир жолу.......................4
Бир жумада бир нече жолу...................5
Күн сайын...................................6

24. Сиздин турак жайыңыз акыркы кынта канча ай жылытылды? ♦

Айдын саны...........................[   ]

25. Сиздин үй чарбанын мүчөлөрү кайсыл жерге жуунушат? ♦

1 Душу бар ванна 1
2 Душ..2
3 Жеке менчиктеги мончо , сауна............3
4 Коомдук мончо, сауна, душ.................4
5 Башка......................................5

26.Үй чарбасында кандай даараткана колдонолат? ♦

1 Канализациянын индивидуалдуу системасы
бар даараткана.............................1
2 Чуңкур казылган даараткана...............2
3 Башка......................................3
4. Даараткана жок...........................4=>суроо 29

27. Даартакана кайда жайгашкан? ♦

Үйдө..1
Короодо....................................2
Көчөдө......................................3

28. Бул даараткананы Сиздин үй чарбасы гана колдонобу, же башкалар менен бирге колдонулатны? ♦

Бир гана үй чарбасы.........................1
Башка үй чарбалары
менен бирге.................................2

29. Сиздин үй чарба акыр-чикирден кантип арылат? ♦

1 Акыр чикир түтүктөрү.....................1
2 Жүк машинасы, контейнер менен жыйноо......2
3 Акыр чикир үймөгүнө таштоо..............3
4 Күйгүзүү...................................4
5 Жерге көмүү................................5

30. Тамак аш даярдоо үчүн эмнени колдоносуз? ♦

1 Примус....................................1
2 Газ баллону бар плита......................2
3 Жердеги электр плитасы.....................3
4 Электроплиткасы............................4
5 Меш, очок..................................5
6 Башка......................................6

31. Коомдук транспорттун жакынкы аялдамасына баруу үчүн Сиздин канча убактыңыз кетет? ♦

5 минуттан аз...............................1
6-15 минут...................................2
16-30 минут.................................3
31-60 минут.................................4
1 сааттан ашык.............................5

|ИНТЕРВЬЮЕР! *32, 33- суроолорду, ижарага алган же ведомстволук турак жайда жашагандарга бербеңиз |*

32. Өзүңүздүн турак жайыңызды канча суммага сатат элеңиз? ♦

Сумма (сом)..................[   ]
Жооп берүүнү
каалабайт..9

33. Эгерде сиз каласаңыз, Өзүңүздүн турак жайыңызды арендага (коммуналдык акысын кошногондо) айына канча суммага берет элеңиз? ♦

Эмереги менен............................[   ]
Эмерексиз....................................[   ]
Жооп берүүнү
каалабайт..9

34. Сиздин кошумча турак жайыңыз барбы? ♦

Ооба..1
Жок...2
                                        ⇓
                          кийинки бөлүм

35. Бул кандай турак жай? ♦

42.1 Өзүнчө квартира......................1
42.2 Айыл жериндеги үй же үйдүн жарымы....2
42.3. Шаар жериндеги үй же үйдүн жарымы.3
42.4 Дача......................................4
42.5 Башка....................................5

V. БӨЛҮМ

УЗАК УБАКЫТ КОЛДОНУЛУУЧУ БУЮМДАРГА ЭЭ БОЛУУ

1. Төмөндө тизмеде көрсөтүлгөн узак убакыт колдонулуучу буюмдардын кайсынысы Сиздин үй чарбаңызда бар? ♦
(жумушчу абалда)

N	Буюмдардын аталышы	Саны
1	А	2
1	Радиоприемник	
2	Музыкалык борбор	
3	Телевизор ак- кара	
4	Телевизор түстүү	
5	Видеомагнитофон	
6	Магнитофон	
7	Видеоплеер	
8	Видеокамера	
9	Сууну электр жылыткыч	
10	Фотоаппарат	
11	Персональный компьютер	
12	Принтер	
13	Кир жуугуч машина жөнөкөй	
14	Кир жуугуч машина автомат	
15	Электр чаң соргуч	
16	Велосипед	
17	Муздаткыч	
18	2-3 камерасы бар муздаткыч	
19	Тоңдургуч	
20	Электрмеш	
21	Микроволновка	
22	Электр жылыткычы	
23	Стол	
24	Стулдар	
25	Керебет	
26	Диван	
27	Кресло	
28	Кресло-керебет	
29	Жүндөн жасалган килем	
30	Жасалма килем	
31	Шырдак	
32	Уюлдук телефон	
33	Кийим үчүн шкаф	
34	Китеп шкафы	
35	Стенка(эмерек)	
36	Китеп текчеси	
37	Эмерек бурчу	
38	Шкафтар	

Авто, мото- техника

N	Буюмдардын аталышы	Саны	Чыгарылган жылы	Сатып алган жылы	Бааланган наркы, сом
1	А	2	3	4	5
1	Жеңил автомобиль				
2	Жүк ташуучу автомобиль				
3	Микроавтобус				
4	Мотоцикл				
5	Мотороллер, мопед				

Талас областы   Маалыматты чогултуучу сырды сактоого кепилдик берет

VI Бөлүм - Айыл-чарба ишмердүүлүгү   6.1. Жер үчүн, мүлзөмөлөрү

1. Сиздин үй-чарбаңызда жериңиз барбы?

| Ооба | ... | 1 |
| Жок | ... | 2 |

2. Сиз өтүүчү жериңизди айыл-чарба иштерине пайдаланасызбы?

| Ооба | ... | 1 |
| Жок | ... | 2 |

3. Эгерде жок болсо, эмне себеп болду

Акча жок	1
Жүгүүчү күчү жок	2
Каналабай	3
Пайдасыз	4
Жокп бертти келебейт	5
Башка себеп	6

Жер ресурстарын пайдалану

Сиздин менчигиңизде жер ресурстарынын кандай түрлөрү бар?

Код	Жер ресурстарынын түрлөрү	Участоктун аянты Эгерде жериңиз мынча табы жок болсо 0 комуру, жериңиз башка табыңс		Сиздин жерге менчигиңизди тастыктаган документиниз барбы?	Жердин баасы кандай?	Сугаттын негизги булагы Канал - 1, Резервуар - 2, Насос - 3, Дарыя - 4, Жаан - 5, Башка - 6	Сиз жериңизди арендага бересизби?	Сиз жериңизди арендага канча бөлүгүн арендага бересиз?	Арендалык төлөмго төлөнго канча акча аласыз? (акыл түрүсү жана/же төлөмнун акчасын көнкөнзө)
		Сотка - 1 Гектар - 2		Ооба - 1 Жок - 2			Ооба - 1 Жок - 2	Сотка - 1 Гектар - 2	
		Саны	Код		сом	код		Саны	Код	сом
1	Сугат жерлер									
2	Кайрака жерлер									
3	Чөп чабык жерлер									
4	Бакчамп-жемиш									
5	Короолору участок									
6	Башка түрлөрү									

13. Жерге арендага аласызбы?

| Ооба | ... | 1 |
| Жок | ... | 2 Кийинки секцияга өтүңүз |

14. Жер ресурстарынын кандай түрүн арендага алгыңыз келет ?

Код	Жер ресурстарынын түрлөрү	Арендага алынган участоктун аянты		Сиз арендага алынган жериңи лоуку-менти барбы?	Сиз жер салынып коюп канча арендага акысына канча төлойсуз? (акыл түрүсү жана/же төлөмнун акысын коюнзо)	Сугаттын негизги булагы Канал - 1, Резервуар - 2, Насос - 3, Дарыя - 4, Жаан - 5, Башка - 6	Сиз арендага алган участкыны сатыш алгыңыз келеби?	Сиз кайсы сумманга участок сатып алууну каалайсыз?
		Сотка - 1 Гектар - 2		Ооба - 1 Жок - 2			Ооба - 1 Жок - 2	
		Саны	Код		сом			сом
1								
2								
3								

Маалыматты чогултууну сырды сактоого кепилдик берет

6.2. Жер ресурстарынын мүнөздөмөлөрү

1. Сиз жер участогуңуз менчиктининзби кепилби?
   - Ооба
   - Жок
   - Билбейм

2. Жер участогуну кепилге/ижарага кайсы аялуу менен?
   - Конурга участогуту сатып алып
   - Мүмкүн болгон участогуту сатып алып
   - Участогуту ижарага алып
   - Оз участогубузду жерим алып
   - башка жол менен (көрсөтүнүз) ......

3. Участогуз уз богуунузду кайсы фирманы жошат?
   - Аралаш
   - Жеке менчикке сатып алуу

4. Кийинки беш жылдыздын ичинде жолго кайча карыавыт жумшоого даярсыз? ...... сом

5. Жакынкы эч жакылын берн жер сатып аллапсынбы? Ооба Жок

6. Кайсы бир менси алган участкыны аласынбы?
   Жер тилкесинин
   Өлчөму
   Баасы

Жердин сапаты

6. Конуу участкыларга салыштырмалуу жердиниздин кандай баалайсыз?

	Жер улугу	Корео участкысы	Баанасы
Эн жакшы	4		
Жакшы	3		
Орточо	2		
Билбейм	1		

7. Сиздин жериниздин кандай проблемалары бар?

	Ооба	Жок	Билбейм
Тузсуздугуту			
Шорлуу			
Саздак			
Ташту			
Суу менси камсыз калуу			
Участкага баруу мүмкүнчүлүгү			

8. Сиздин участогуз уйрутуте чейинки аралык канчал?

	Жер улугу	Баанасы
1 жакшт аз	1	
1 дни 5 км ч. н	2	
5 дни 10 км	3	
10 ашкен жогору	4	

9. Сиздин участокко чейинин аралык проблема жаратабы?
   - Ооба
   - Жок
   - Билбейм

10. Сугат системасы кандай баалайсыз?
    - Эн жакшы 4
    - Жакшы 3
    - Орточо 2
    - Жаман 1
    - Билбейм 0

11. Сугат уодуруда сизге проблема болобу?
    - Ооба
    - Жок

12. Сугат үчүн канча толейсиз, сомга? ...... сом

13. Бир сезонда канча алуу сугарасыз ......
    - Айдоо жер
    - Корео участогу

14. Жердин сапатын жакшыртуу боюнча чаралады көрүсүз жа?
    - Ооба
    - Жок

15. Кандай чараларды көрсүз?
    - Минералдык жер семирткич колдоном
    - Органикалык жер семирткич колдоном
    - Айланыш жер алмаштыром
    - Дрекаждык ишгерузди аргулучум
    - Жер эмгутынын текшылем
    - Тосмо коюм

10

145

Талас областы   Маалыматты чогултууну сырды сактоого кепилдик берет

6.3. Айыл-чарба өсүмдүктөрүн өстүрүү жана түшүмүн пайдалануу

Сиздин үй-бүлөңүз айыл чарба өсүмдүктөрүн өстүрөбү?   Ооба   1
                                                         Жок   2 Кийинки секцияга өтүңүз

1. Негизги участоктогу өсүмдүктөр

Бышкан түшүмдү көрсөтүңүз, бышылбаган жок болсо, откон жылдагыны көрсөтүңүз

Код	Аталышы	Аянт (сотка - 1, га - 2)		Түшүм	Сиз канча продукция саттыңыз?	Табылган каражат	Урунга калганы	Заявкечтерден, куткарылгандар , ж.б. тартуан жоготуулар	Сиздерди үй-бүлөө азык түлүктү тамакка колдонобу?	Келечекте керектөө же сатуу үчүн иштелип чыкты	Сактоо үчүн калтырылды
		Саны	Код	кг	кг	сом	кг	кг	кг	кг	кг
1											
2											
3											
4											

Калган продукциянын канчасын сиз ......

Код	Аталышы		Белек катарында бердиңиз	Алмаштырдыңыз	Алмашам дегенге алдыңыз	Карызга бердиңиз
		кг	кг	кг	кг	кг
	Аталышы			Аталышы		

2. Короодо жана башка кошумча участоктон өстүрүлгөн азык-түлүк

Код	Аталышы	Түшүм	Тамак-ашка колдонулду	Сатылды		Иштелип чыкты	Сактоого калтырылды
		кг	кг	кг	сом	кг	кг
1							
2							
3							
4							
5							
6							
7							
8							
9							

11

146

6.4. Айыл- чарба иштерди жүргүзүү үчүн сатып алынган каражаттар

1. А.Ч. иштерди жүргүзүү үчүн сатып алынчубу?

Камсыздоочунун коду

а Урөн	Базардан Алыстоочки камсыздоочудан	1 Таманынтырдан
б Минералдык жер семирткич		2 Баштапкы (көрсөтүлүп)
в Органикалык жер семирткич		3 Сатып алынбайбат ... 5
г Өрчөө материалдары / мелюк		4

2. Нысаьга сатып алынчубу   3. Канча бөлүгүн (%) насыяга сатып алынган?

Ообба

Жок

14. Техниканы иштетиш аткаруга каражат жумшалынчубу?

Механизациянын иштери

			Сумма сом	х харалма училге тыжымдыру	Калгал карыл
а	Тузаки				
б	Айдоо				
в	Себуу				
г	Тыгыздоо жыйноо				
	Башкалар				

17. Силү үч чарбасынын башкы калдай саргытоолоорду жумшалынылат?

а	Жер салыгы	
б	Соц Фонд	
в	Башкы салыктар	
	Банкга карыжат	

Талас областы   Маалыматты чогултуучу сырды сактоого кепилдик берет

6.5. Акыркы 12 айдын ичинде сиз өндүргөн тамак-аш азыктары

1. Акыркы 12 айдын ичинде жеке өзүңүздүн же сатып алган продукциядан тамак-аш азыктарын өндүрүүчүбү?
Ооба 1
Жок .... 2 Кийинки секцияга өтүңүз

2. Сиздин үй-чарбаңыздын кайсы мүчөсү тамак-аш азыктарын чыгарууга тартылган?
Эгерде үй-бүлөңүздү бир нече мүчөсү тамак-аш азыктарын берген үч адамдын колун коюңуз
Кол [ ] Кол [ ] Кол [ ]

Кол	Аталышы	Акыркы 12 айдын ичинде сиздин үй-бүлө өзүнөрдүн же сатып алган продукциядан тамак-аш азыктарын өндүрүнүзбү? Ооба - 1 Жок- 2 -> кийинки Азык	Акыркы жылы канча азыктарды өндүрдүнөр?	Акыркы 12 айдын ичинде канча тамак-азык канча азык-түлүк колдондуңар?	Акыркы 12 айдын канча ай бул азыкты тамактанууга колдонондуңсуз?	Акыркы 12 айдын ичинде сиз канча азык-түлүктү азыктарда жардам жана белек катарында берднниз?	Азык-түлүктүн канчасы сакталого калат?	Акыркы 12 айдын ичинде канча азык-түлүк саттыныз?	Акыркы 12 айдын ичинде азык-түлүктү сатуудан сиздин үй-чарбаңыз канча акча тапты?	Азык-түлүктү өндүрүү үчүн канча акча жумшадыңы
						Эгерде 0 болсо, кийинки 7-суроого өтүнүз	Эгерде берилбесеңиз, 0 жазыныз	Эгерде калбаса, 0 жазыныз	Эгерде 0 болсо, кийинки 11 суроого өтүнүз		
			саны	айдын саны	саны	саны	саны	саны	сом	сом
1	Кант										
2	Консервацияланган жашылча										
3	Татымал										
4	Ун										
5	Шире										
6	Компот										
7	Кургатылган жемиш										
8											
9											
10											
11											
12											
13											

13

Маалыматты чогултуучу сырды сактоого кепилдик берет

6.6. Мал чарбасы, мал чарба жандыктары

1. Сиздин үй чарбаңызда мал чарба жандыктары барбы?
   Ооба ....
   Жок 1
2. Сиздин үй бүлөңүздүн кайсы мүчөсү мал багат? ..... 2 кийинки бөлүмгө өтүңүз

Эгерде үч адамдан ашык болсо, эң көп ээлеген үчүнчү жазыңыз

код [   ]   код [   ]

N п/п	A	Бир жаш мурда канча мал жандыгыңыз бар эле? саны	Азыркы жылдыгы ичиндеде канча мал сатгыңыз? саны	Туруган пайдаланып суммасы канча? сом	Азыркы жылдыгы канча мал жандык сатып алдыңыз? саны	Мал үчүн канча сумма төлөдүңүз? сом	Азыркы жылы мал жандыгы башка канчага кошуу боюнча (болонуз берилиши)? саны	Азыркы жылы мал жандыгы ичине сиз канча малды тамак аша колдонуунуз? саны	Азыркы жылдын ичине сиз канча малды ичкиликте жылдын ашуу колдонуунуз же уурдатгыныз? саны	Азыркы жылы сиз канча малды менишден акрутуунуз же уурдатгыныз? саны	Азыркы жылга сиз канча малды өлсөнгө берилиши? саны	Азар сиздин канча малыныз бар? саны	Бүгүнкү күндө сиздин малыныздын баалы канча? сом
		3	4	5	6	7	8	9	10	11	12	13
1	Үй												
2	торпок жана музоолор												
3	Бука жана өгүз												
4	Бир жашта кочкор												
5	Чочко												
6	Торпой												
7	Кой												
8	Козу												
9	Эчки												
10	Улак												
11	Жылкы жана бээ												
12	Кулун												
13	Кунаттуу												
14	Эшек												
15	Башка жаныбарлар												

14

Талас облаты   Маалыматты чогултуучу сырды сактоого кепилдик берет

6.7.Мал чарба азыктары

1. Сиз керектоо, сатуу, белек кылуу үчүн кандайлар бир мал чарба азыктарын өндүрүүчү беле?
   Ооба 1
   Жок ....
2. Силдин үй бүлөнүздүн кайсы мүчөсү мал багат?
   2 кийинки болуучу өтүнүч
   Үч адамдын атык болсо, эн көп ишлеген үчөнү жазыныз

N п/п	А	Бир жылда канча азык өндүрдүңүз? саны	Бул азыкты бир жылдын канча айы керектейсиз? айдын саны	Адатта сиз бир айда канча азык түрүк керектейсиз? саны	Сиз башка адамдарга бекер азык түрүк бересизби? саны	Адатта сиз үйдө канча азык түрүкту үйдө сактайсыз? саны	Акыркы 12 айдын ичинде канча азык түрүкту сактыныз? саны	Азык түрүкту сатууда канча каражат таптыныз? сом	Азык түрүкту өндүрүүгө Сиздин канча каражатыныз кетти? сом
1	Койдун эти								
2	Уйдун эти								
3	Кшнун эти								
4	Жылкырка								
5	Сүт								
6	Нетизги эмес азыктар								
7	Жун, кийиз								
8	Тери								
9	Бал								
10	Балык								
11	Башка азыктар								
12									

15

Талас областы　　Маалыматты чогултуучу сырды сактоого кепилдик берет

6.8. Мал жанымалстарды каароо үчүн чыгашалар

1. Ветеринардык чыгашалар

№	Чыгашалар	Малдын саны	Бир баш малга жасалган чыгашалоон	Жылдык суммасы
1	коойлор			
2	Пул медетүлүп мал			
3	айлалгар			
4	контуттулар			
5	башка(корректтуу)			

2. Малдын жайышынын чыгашалары

№		Малдын саны	Бир баш малга жасалган чыгашалоон	Жайыкка чыгарган айлын саны	Жылдык суммасы
1	Айлалык жайыты				
2	айлар				
3	аре медетүлүп мал				
4	жаамалар				
5	алмаха жайышттар				
6	коойлор				
7	аре медетүлүп мал				
8	жаамалар				

3. Токт сатып алуута кеткен каражаттар

№		Малдын саны	Бир баш малга жасалган чыгашалоон	Жылдык суммасы
1	Дан			
2	Чеп			
3	Тут			
4	Аралаш токт			
5				
6				
7				
8				

4. Кам чыгашалар

№	Элмекке актымакшы	Жылдык суммасы
1		
2		
3		
4		
5		
6		
7		
8		

5. Сизге кжансайкы бир айдыл чарба техникасы же гартуучу уата барбы?

Ооба......................................1

Жок..2 12де

6. Сизге кжансайкы бир айдыл чарба техникасы же гартуучу уата барбы, а.т салдан кечгүп имилеп, же аренада алгаксылбы?

N п/п	Аталышы	Саны	Жекге мезгилк бир үч колдонот-2		Аренда бир үч колдонот-1 колдонот бир нече үй чарб. колдонот-2
	А		бир үч колдонот-1 бир нече үч чарб колдонот-2		
			2	3	4
1	Жумушчу аттар				
2	Жумушчу кочкор(шнектер)				
3	А табы				
4	Суу насосу				
5	Урон ставич				
6	Трактор				
7	Мотоблок				
8	Мини-трактор				
9	Чеп чаппыч машина				
10	Комбайн				
11	А.ч ишинакасы үчүн селел				
12	Телегелерук чиркемес				
13	Башкасы (көрсөтүлгү)				

16

Талас областы   Маалыматты чогултуучу сырды сактоого кепилдик берет

VII БӨЛҮМ – АЗЫК – ТҮЛҮК ЧЫГАШАЛАР

1. Акыркы 14 күндүн ичинде Сиздин үй чарбаңыздын мүчөлөрүнүн кимдир бирөөсү жок болду беле?

Ооба   -1
Жок   (2) →   4 суроо

2. Акыркы 14 күнү ичинде үй чарба мүчөлөрүнүн кимиси 2 күн жок болду?

НК (   ) (   ) (   )

3. Көрсөтүлгөн үй чарба мүчөлөрү акыркы 14 күндү канча күнү болгон эмес?

Күнү (   ) (   ) (   )   код (   )

12. Сиз насыяга сатык аласызбы?
    Ооба   -1
    Жок   -2   (   ) сом

13. Акыркы 14 күндүн ичинде Сиз канча насыяга сатык алдыңыз?

14. Сиз башынга азыкка же товарга алмашып сатык жүргүздүңүзбү?
    Ооба   -1
    Жок   -2

15. Акыркы 14 күндүн ичинде канча суммага бартердик алмашуу жүргүздүңүз?
    (   ) сом

№	Аталышы	3 Акыркы 12 айдын ичинде көрсөтүлгөн азыктарды колдондуңузбу? Ооба -1 Жок 2 Код	4 Акыркы 12 айдын канча айы Сиз көрсөтүлгөн азыктарды сатып алдыңыз? Айдын күнү	5 Аталган азыктарды канча жолу сатып алдыңыз? 1 - күн сайын, 2 - жума сайын, 3 - ай сайын, 4 - квартал сайын, 5 - жыл сайын Код	6 Адатта Сиз канчаны сатып аласыз? Грамм - 1 Килограмм- 2 Литр- 3 Бирдик- 4 Саны Код	7 Ортоңо сатыкта көрсөтүлгөн азыктарга Сиз канча жумшайсыз? сом	8 Адатта көрсөтүлгөн азыктарды кайдан сатып аласыз? 1 - базар, 2 - дүкөн, 3 - чакан-рынок, 4 - жеке адамдардан 5 - башкасы Код
1	Ак нан						
2	Токоч						
3	Буудай уну						
4	Күрүч						
5	Макарон, лапша, кесме						
6	Картошка						
7	Сабиз						
8	Пияз						
9	Помидору						
10	Бадыраң						
11	Сарымсак						
12	Болгар калемпири						
13	Ачуу калемпир						
14	Капуста						
15	Башка жашылчалар						

17

№	Аталышы	3 Акыркы 12 айдын ичинде керетүлгөн азыктарды колдоносузбу? Ооба -1 Жок 2 Код	4 Акыркы 12 айдын канча айы Сиз керетүлгөн азыктарды сатып алдыңыз? Айдын күнү	5 Аталган азыктарды канча жолу сатып алдыңыз? 1 - күн сайын, 2 - жума сайын, 3 - ай сайын, 4 - квартал сайын, 5 - жыл сайын Код	6 Адатта Сиз канчаны сатып аласыз? Грамм - 1 Килограмм - 2 Литр - 3 Бирдик - 4 Саны Код	7 Орточо сатыкта керетүлгөн азыктарга Сиз канча жумшайсыз? сом	8 Адатта көрсөтүлгөн азыктарды кайдан сатып аласыз? 1 - базар, 2 - дүкөн, 3 - чакан-рынок, 4 - жеке адамдардан, 5 - башкасы Код
17	Алма						
18	Алмурут						
19	Тропикалык жемиштер						
20	Башка жемиштер						
21	Сухофрукты						
22	Кургатылган жемиштер						
23	Дарбыз						
24	Коон						
25	Томат-паста						
26	Баңдар аягы						
27	Консерв. жашылча(балдыран)						
28	Бал						
29	Жаңгак и чемичке						
30	кой эти						
31	Уй эти						
32	Жылкынын эти						
33	Ич эт уйдуку						
34	Жылкынын (чучук, казы, карта)						
35	Колбаса жана сосиска						
36	Тоокун ж.б. канаттуунун эти						
37	Жаны балык						
38	Консерваланган балык						
39	Консерваланган эт						

18

153

Талас областы   Маалыматты чогултуучу сырды сактоого кепилдик берет

№	Аталышы	3 Акыркы 12 айдын ичинде көрсөтүлгөн азыктарды колдондунузбу? Ооба -1 Жок 2 Код	4 Акыркы 12 айдын канча айы Сиз көрсөтүлгөн азыктарды сатып алдыныз? Айдын күнү	5 Аталган азыктарды канча жолу сатып алдыныз? 1 - күн сайын, 2 - жума сайын, 3 - ай сайын 4 - квартал сайын 5 - жыл сайын Код	6 Адатта Сиз канчаны сатып аласыз? Грамм - 1 Килограмм - 2 Литр- 3 Бирдик- 4 Саны Код	7 Ортоно сатыкта көрсөтүлгөн азыктарга Сиз канча жумшайсыз? сом	8 Адатта көрсөтүлгөн азыктарды кайдан сатып аласыз? 1 - базар, 2 - дүкөн, 3 - чакан-рынок 4 - жеке адамдардан 5 - башкасы Код
40	Жумуртка						
41	Сүт						
42	Кефир						
43	Айран						
44	Йогурт						
45	Маске май						
46	Голланд сыры						
47	Маргарин						
48	Майонез						
49	Өсүмдук майы						
50	Мал майы						
51	Чай						
52	Кофе						
53	Алкоголдук ичимдик						
54	Максым, бозо						
55	Шире						
56	Газдалган суусундуктар						
57	Кумшекер						
58	Туз						
59	Калемпир ж.б. Кошумчалар						
60	Кондитердик азыктар						
61	Сигаретер						

19

VIII БӨЛҮМ- АЗЫК ТҮЛҮК ЭМЕС ТОВАРЛАР ЖАНА КЫЗМАТТАР

1. АЗЫК ТҮЛҮК ЭМЕС ТОВАРЛАР САТЫП АЛУУГА КЕТКЕН ЧЫГАШАЛАР

Келиңиз чыгашалардын бардык түрлөрү жөнүндө сүйлөшөлү

(эгерде мындай сарптоолор болсо)

№/п	Чыгашалардын аталышы	Акыркы үч айдын ичинде (сом)		№		
1	А	5		11	Зергер буюмдары	
1	Кийим , ички кийим			12	Транспорт каражаттары	
2	Баш кийимдер			13	Тиричилик электр буюмдары	
3	Жоолук, бир байлам жоолук, шарфтар			14	Ашкана идиштери	
4	Кол кап, мээлей			15	Эмерек	
5	Байпактар			16	Үй үчүн текстиль буюмдары	
6	Бут кийим			17	Айыл чарба шаймандары	
7	Кездемелер			18	Курулуш материалы жана сантехникалык жабдуулар	
8	Кийимге башка кошумча жасалгалар			19	Башка азык түлүк эмес товарлар	
9	Булгаары галантереясы			20	Баардыгы (сом)	
10	Теле, радио ж.б. аппаратура					

2. ТУРАК ЖАЙ-КОММУНАЛДЫК ЧЫГАШАЛАР

Азыр мен Сиз жашаган турак жайдын коммуналдык чыгашаларын кенен жазайын дегем. **Акыркы үч айдын ичинде отун сатып алууга байланыштуу чыгашалар болдубу?** (буга сатып алууга жумшалган каражат жана белек катарында алынганы да кошулат)

№/п	Отундун түрү	Өлчө бирд. 1-куб.м 2-кг 3-ц 4-т	Акыркы жылы канча отун сатылып алынды?	Акыркы жылдагы отундун баасы? (сом)	Акыркы жылы жылытууга канча отун кетти?	Канчасы н өзүнүз таптыны з (өзүнүз чогултту нуз)	Канчасын белек катарында алдыныз?
1	А	4	5	6	7	8	9
1	Жыгач отун						
2	Тезек						
3	Көмүр						
4	Мазут						
5	Башкасы						

20

155

2. Жеңилдиктерди эсептеп жана эсептебей коммуналдык кызматтарга жана электр энергиясына төлөмдөр

(Менчик үйлөрдө жана квартираларда жашаган үйбүлөлөр жооп беришет)

№/ п	Кызматтын түрлөрү	Сиз төмөндөгү кызматтын түрлөрүнөн пайдаланасызбы? Ооба....1 Жок...2⇒ кийин. Кызмат	Ай сайын кызматтарга канча төлөгөнүнүздү көрсөтүнүз (сом)	
			Жай мезгилинде	Кыш мезгилинде
1	А	3	6	7
1	Электр энергиясы			
2	Муздак суу жана канализация			
3	Акыр чикирди алгандыгы үчүн			
4	Башка коммуналдык кызматтар (кайсылар көрсөтүнүз)			

3. САЛАМАТТЫКТЫ САКТООГО ЧЫГАШАЛАР

1. Акыркы үч айдын ичинде Сиздин үй чарбаңыздан кимдир бирөө медициналык жабдууларды (көз айнек, угуучу аппаратты, балдакты ж.б.), дары дармекти, чөнтөрдү сатып алууга, ошондой эле медициналык мекемеге кайрылгандыгы, жатып дарылынгандыгы үчүн канча акча жумшады?

Ооба...1

Жок..2 ⇒ 4 бөлүкчө

Жооп бергим келбейт...9 ⇒ 4 бөлүкчө

2. Бул акча эмнеге жумшалды? (сом)

№/ п	Чыгашанын түрлөрү	Үй бүлө мүчөсүнүн коду	Акыркы үч айдын ичинде (сом)
1	А	1	3
1	Медициналык жабдуулар (балдак, майып үчүн коляскалар, грелка, контрацептив, контактык линзалар ж.б.)		
2	Дары дармек		
3	Догдур кабыл алуусунда болуу		
4	Тиш догдур кабыл алуусунда болуу		
5	Лабораториялык анализдер, УЗИ, рентген, диагностикалык изилдөө, физиотерапия, массаж, ийне сайуу ж.б.		
6	Догдур жана медкызматкерге расмий төлөмдөр		

21

Талас областы   Маалыматты чогултуучу сырды сактоого кепилдик берет

7	Тамак ашка		
8	Дары дармек		
9	Доттурга жана Медкызматкерлерге расмий эмес төлөмдөр (белектерди кошкондо)		
10	Лабораториялк анализдер, УЗИ, рентген, физиотерапия ж.б.		
11	Битттер, шприцтер, ийне сайуу, шейшептер		
12	Башка чыгашалар (кенен көрсөтүнүз)		

4. ТРАНСПОРТКО ЖУМШАЛГАН КАРАЖАТТР

Мен транспортко кеткен чыгашалар жөнүндө сүйлөшөйүн детем.

1. Сиздин үй бүлөнүз шаардан сырткаркы же шаар аралык каттамдардагыжүргүнчүлөр транспортунун кызматынан пайдаланды беле? (окууга байланыштуу коомдук шаардык транспортту пайдаланганды кошпогондо)

Ооба...1

Жок...2 ⇒ бөлүкчө 5

Жооп бергим келбейт.............................…..........9 ⇒ бөлүкчө 5

2. Сиздин үй бүлөнүз транспорттун кайсы түрүн пайдаланат, жана бул кызматка канча акы төлөйсүз? ₩(сом)

№ /п	Транспорттун түрү		Акыркы үч айда (сом)	Үй бүлө мүчөсүнүн кодун көрсөтүнүз
1	А		5	
1	Автобус	Республиканын чегинде		
2		Республиканын чегинен тышкары		
3	Такси	Республиканын чегинде		
4		Республиканын чегинен тышкары		
5	Темир-жолу	Республиканын чегинде		
6		Республиканын чегинен тышкары		
7	Аба жолдору	Республиканын чегинде		
8		Республиканын чегинен тышкары		

22

157

5. БИЛИМ АЛУУГА ЧЫГАШАЛАР

1. Сиз балдарды багууга жана аларды мектепке чейинки мекемелерге берүү үчүн кандайдыр бир каражат жумшадыңызбы?

Ооба...1

Жок...2 ⇒ суроо 3

2. Акчаны эмнеге корттунуз? (сом)

код	Тейлөөнүн аталышы	Акыркы жылга (сом)
1	А	4
1	Мектепке чейинки мекемеге төлөм	
2	Мектепке чейинке мекемеге керек буюмдарды сатып алуу	
3	Бала багуучу кызматы	
4	Расмий эмес чыгашалар (белектер)	
5	Башка төлөмдөр (кенен көрсөтүнүз)	

3. Акыркы жылдын ичинде Сиздин үй чарбаңыз билим алууга каражат жумшады беле?

Ооба...1

Жок...2 ⇒ бөлүкчө 6

Жооп бергим келбейт..9 ⇒ бөлүкчө 6

4. Окууга канча акча жумшадыңыз? (өткөн окуу жылына кеткен сумманы сом менен көрсөтүнүз)

Окуу жайы			Башталгач окуу жайында	Орто окуу жайында	Атайын орто окуу жайында	Жогорку окуу жайында	Курстарда
Сиз төмөндөгү кызматтардан	Ооба— 1	Жок -2					
Окуу үчүн төлөм							
Окуу китептери жана канцелярдык							
Мектеп, спорт кийимдери							
Китепканадан пайдалануу							
Репетитор жалдоо и кошумча сабак							
Транспорттук чыгашалар							
Мектептин оңдоосуна							
Белек жана төлөмдөр үчүн чыгашалар							
Башка чыгашалар							
Чыгашалардын жалпы суммасы							

23

158

6. БАШКА ЧЫГАШАЛАР

1. Сидин төмөндөгү кандайдыр бир чыгашаларды жасады беле? ⇓(сом)

Код	Чыгашалардын түрлөрү	Акыркы Үч айдын ичинде (сом)
1	А	4
1	Санаториялык-курорттук жана туристтик-экскурсиялык кызматтар	
2	Камсыздоонун бардык түрлөрү боюнча төлөмдөр	
3	Транспорт каражат ээлеринин чогулган кражат	
4	Байланыш, интернет, радио, телефон (квартиралыкты кошкондо,) кызматтары	
5	Мобильдик байланыш кызматы	
6	Укуктук мүнөздөгү кызматтар	
7	Ритуалдык кызматтар (үйлөнүү тою, маркумду узатуу ж.б.)	
8	Туугандарга, досторго, тааныштарга берилүүчү салтанатка, юбилейге, үйлөнүү тоюна, маркумдуу узатууга байланыштуу белектер	
9	Башка кызматтар жана чыгашалар (кенен көрсөтүнүзподробнее)	

2. Сиздин үй бүлөнүздөн кимдир бирөө алимент төлөйбү?

Ооба...

Жок..1 ⇒ 3 суроо

3. Төлөнүүчү алименттин суммасы кандай? ⇓

	Акыркы квартал үчүн
Сумма (сом)	

3. Сиздин үй бүлөнүз бирге жашабаган туугандарга же досторго жардам берди беле?

Ооба...1

Жок..2 ⇒ кийинки бөлүкчө

4. Бул жардам кандай түрдө берилди? Эгерде акча формасында болсо, сумманы, эгерде акча эмес түрдө болсо, болжолдуу баасын көрсөтүүгө аракет жасаңыз. ⇓

Код	Жардамдын түрлөрү	Акыркы квартал үчүн
1	А	2
1	Акча	
2	Сиз өндүргөн тамак аш азыктары	
3	Сиз сатып алган тамак аш азыктары	
4	Кийим, бут кийим	
5	Дары дармек	
6	Башкалары(кенен жазыңыз)	

24

159

Талас областы   Маалыматты чогултуучу сырды сактоого кепилдик берет

IX БӨЛҮМ - Үйдөн сырткары тамактануу, тамак аш азыктарын башка жактан алуу

1. Сиздин үй бүлөө мүчүнүздүн кимдир бирөөсу сырттан тамактанабы?

Ооба ... 1
Жок ... 2

кийинки суроого өтүнүз 2

№		Бир айда канча жолу	Бир жолукута ортоочу сумма, сом	Жалпы сумма, сом
	Үй бүлө мүчөсүнүн коду ( )			
1	Эртең мененки тамак			
2	Түшкү тамак			
3	Кечки тамак			
4				

№		Бир айда канча жолу	Бир жолукута ортоочу сумма, сом	Жалпы сумма, сом
	Үй бүлө мүчөсүнүн коду ( )			
1	Эртең мененки тамак			
2	Түшкү тамак			
3	Кечки тамак			
4				

№		Бир айда канча жолу	Бир жолукута ортоочу сумма, сом	Жалпы сумма, сом
	Үй бүлө мүчөсүнүн коду ( )			
1	Эртең мененки тамак			
2	Түшкү тамак			
3	Кечки тамак			
4				

2. Акыркы жолу Сиз .досторлон, туугандардан же гуманитардык уюмдардан бекер тамак аш алдыныз беле?

Ооба ... 1
Жок ... кийинки бөлүкчөгө өтүнүз

3. Акыркы 12 айда Сиз канча жолу тамак аш алдыныз?

Дата	Тизме

4. Сиз алган тамак ашпган бассы кандай? сом

5. Кийинки 12 айда .досторлон, туугандардан, гум. уюмдардан тамак аш түрундө жардам алууга ниеттенибсизби?

Ооба ... 1
Жок ... 2
кийинки бөлүкчөгө өтүнүз

6. Жардам откен жылдын көлөмундө болот деп күтосүзбу?

Ооба ... 1
Жок ... 2
Билбейм
кийинки секцинага өтүнүз

Талас областы   Маалыматты чогултуучу сырды сактоого кепилдик берет

X БӨЛҮМ - Карыздар жана топтогон акча

1. Сиз акча каражаттарын досторго, кошуналарга, туугандарга же бизнес өнөктөрүнө карызга бересизби?
   Ооба   1
   Жок   2 3 суроого өтүңүз

2. Азыркы учурда Сизге канча акча карыз?
   сумма _____ сом

3. Сиз же үй-бүлөңүздүн мүчөсү карызга канчалык бир товар алдыңыздар беле?
   Ооба   1
   Жок   2 айынан 6 суроого өтүңүз

4. Акыркы 12 айда сиз сатып алган азыктын жана кызматтын жалпы суммасы кандай?
   сумма _____ сом

5. Азыркы учурда карызыңыздын жалпы саамасы канча?
   сумма _____ сом

6. Акыркы 12 айдын ичинде Сиз же үй-бүлө мүчөлөрү банктардан, микрокредиттик уюмдардан/компаниялардан же асса адамдардан, туугандардан жана досторлон алган карыздардын, насыялардын суммасы канча?
   Ооба   1
   Жок   2 10.1. бөлүмүнө өтүңүз

7. Сиз акча сиздин үй-бүлө мүчөлөрү али төлөм боло элек канча насыя жана карыз алдыңыздар?
   Саны (_____)

8. Насыянын жалпы суммасы канча?   сумма (_____) сом

10.1. Топтолгон акчалар и депозиттер

1. Сиздин жана сиздин үй-бүлөңүздүн топтолгон акчасы же банкта каражаты барбы?
   Ооба   1
   Жок   2
   Бар болсо,банктын карсоотуңуз

2. Азыркы учурда бааш хесебинизде канча акчаңыз бар?
   сумма (_____) сом

3. Акча топтой баштаган болсоңуздар убакытты көрсөтүңүз?

4. Банктын шарттары силер киришүүсүбу, эгерде жашаса, кайсынысыла көрсөтүңүз
   Ооба   1
   Жок   2

26

Талас областы   Маалыматты чогултуучу сырды сактоого кепилдик берет

10.2. Үй-чарба мүчөлөрү алган карыздар жана насыялар

Насыялар жана карыздар	Сизге насыя берген ким? Тууган...1 сулоо...2 Мамлекеттик банк...3 Коммерциялык банк...4 Микрокредит уюм...5 Башкалар... 6	Насыянын суммасы канча? сом	Насыя алуунун датасы айы / жылы	Насыя алуунун максаты? А ч. эмнөрсачтуу ...1 Жеке сатып алуу ...2 Жакшыртылган сатып алуу... 3 Бизнес жүргүзүү ...4 Курулуш же мүлчөн сатып алуу ...5 Окууга...6 Башка алуу...7 Башка...8 код	Сиздин пайыздык/карыз атайбир жылдык пайыз канлай коюптай? консенген акча	Сиз же үй-бүлө мүчөлөрү алган насыя/карыздыкс пайдалык эсеби болуптубу? болуптубу? Ооба...1 Жок...2 кийинки суроочогулуу 8	Капиталдыкс эмисти койдуну? Турук жай...1 Жер...2 Бакоо буюмдар...3 Жеке буюмдар...4 Башка...5 код
А	код	сом	айы / жылы	код	консенген акча	код
Биринчи насыя (карыз)						
Экинчи насыя (карыз)						
Үчүнчү насыя (карыз)						
Төртүнчү насыя (карыз)						

Насыялар жана карыздар	Сиз насыяңкарыз боюнча төлөмдөрдү жүргүзүп атасыцбы? Ооба...1 Жок...2 11 суроого өтүнүз	Карызды төлөмдү жана карыз жолу төлөм жүргүзүлүп? күн сайын...1 ай сайын...2 карталда сайын...3 жарым жылда бир жолу...4 жыл сайын...5 бир жолу...6	Акыркы 12 айдын ичинде толук төлөм жүргүзүлдубу? Ооба...1 Жок...2 11 суроого өтүнүз	Бугунку күнге насыяңын канчасы калды?	Насыяны акыркы чейин төлөп бүтүү-өгүгө таш датасы жок болсо, 0 коюнуз
А	сом	мөзгилдүгүлүг	сом	сом	айы / жылы
Биринчи насыя (карыз)					
Экинчи насыя (карыз)					
Үчүнчү насыя (карыз)					
Төртүнчү насыя (карыз)					

27

162

XI бөлүм- Айыл чарбасына байланышпаган ишмердүүлүк

1. Сиз же үй чарбасынын мучөсү өз алдынча ишке эмгектенеби?

    Ооба   ...   1   мучөнүн кодун көрсөтүңүз

    Жок   ...   2   үй чарбасы

2. Сиздин ишиңизди көрсөтүңүз   У.ч. мүчөсүнүн коду

    а   тамак аш азыктарын сатуу

    б   тамак аш эмес товарларды сатуу

    в   транспорттук кызматтар

    г   А.ч. азыктарын иштеп чыгаруу

    д   жалданган сезондук жумушчу

    е   башкасы (көрсөтүңүз)

3. Сиз өз алдынча иштейсизби же жалданма кызматкерсизби?

                        У.ч. мүчөсүнүн коду

    а   өз алдымча иштейм

    б   жалданып иштейм   **8 суроого өтүңүз**

4. Акыркы 12 айда өзүнүздүн ишмердүүлүктөн тапкан каражатты көрсөтүңүз?

айдын саны	
ай сайын тапкан киреше (чыгашаны эсентебегенде)	
жалпы киреше	

5. Эгерде өз алдынча иштесеңиз, Сиз кошумча кызматкерлерди жалдайсызбы?

    Ооба   ...   1

    Жок   ...   2   **7 суроого өтүңүз**

6. Акыркы 12 айда канча ай аларды жалдадыңыз жана ортоочо айлык акысы канча?

Саны	
Ортоочо айлык акысы	
Айлардын саны	
Чыгашалардын жалпы суммасы	

7. Бизнести жүргүзүү үчүн ай сайын кеткен чыгашаларды көрсөтүңүз

Патентке төлөм	
Жабдуунун арендасына	
Соода аянтынын арендасы үчүн	
Техниканы, автомобилди ондоо	
Башка чыгашалар	
Чыгашалардын жалпы суммасы	

11.2. Эмгек миграция

9. Сиздин үй бүлөңүздө сырткы мигрант барбы?

    Ооба   Жок

10. Кайсы жылы/айы үй чарбадан кетти?

    жылы   ай

11. Кайсы өлкөдө жана негизги ишмердиги кайсы чойродо?

12. Акыркы 6 айдын ичинде Сиздин үй бүлөңүздүн мүчөсү болгон мигрант тарабынан которулган акчаны алдыңызбы?

    Ооба   Жок

13. Акыркы 6 айдын ичинде баардык үй чарба мүчөлөрү канча акча которушту?

    сумма

8. Эгерде жалдансаңыз канча ай иштедиңиз жана акыркы 12 айда ишиңиз үчүн канча алдыңыз?

айлардын саны	
айлык эмгек акы	
эмгек акынын суммасы	

28

English Version

Talas Oblast *Confidentiality is guaranteed by the recipient of this information*

SELECTIVE HOUSEHOLD SURVEY

LOCATION OF HOUSEHOLD _____ (NAME OF SETTLEMENT, STREET, HOUSE)

Code of HH	
Code of interviewer	
Code of results	10 - completed interview, 11 - interview is not completed, 20 - flat/house is not occupied, 21 - the house has been demolished/no longer exists, 22 - It was converted into a commercial enterprise, 30 - could not contact with the dwellers, 31 - - they refused to answer, 40 - change of place of dwelling, 41 - - other

Actual date of carrying out of interview

day | _ | _ | month | _ | _ | 2011

Time of interviewing

Start: | | | hours | | | minutes

End: | | | hours | | | minutes

- 1 -

164

SELECTIVE HOUSEHOLD SURVEY

Section I. **Household Control Card**

LOCATION OF HOUSEHOLD _____

Code of HH

Code of interviewer

(NAME OF SETTLEMENT)

#	1. First name and Patronymic *(Interviewer make list of all residents of HH before to turn to q. 2-13)*	2. Gender male- **1** female- **2**	3. Family Head Relationship: Head.............1 Husband/wife.............2 Son/daughter.............3 Son-/daughter-in-law.....4 Father/mother.............5 Husb./wife parents and h....6 Brother/sister.............7 Grandson/daughter.........8 Grandmother/father.........9 Nephew/niece............10 Brother/sister of husband/wife............11 Other relatives............12 Servant...........13 Tenant............14 Others............15	4. Interviewer: Is (NAME) a member of the household? Yes....1 No....2	5. Date of Birth For example, 01.05.1995			6. Age in completed years	7. Marital Status: Married...........1 Live together but not married.............2 Divorced.............3 Live separately but not divorced.........4 Widow/widow...5 Never has been married....6 Up to 13 years. 98	8. Nationality: Kyrgyz..........1 Russian.........2 Ukrainian......3 Uzbek..........4 Kazakh.........5 Belorussian...6 Tajik...........7 Tatar...........8 Dungan........9 Others........10
					Day	Month	Year			
1										
2										
3										
4										
5										
6										
7										
8										
9										
10										
11										
12										

- 2 -

165

Talas Oblast

Confidentiality is guaranteed by the recipient of this information

#	1. First name and Patronymic *(Interviewer make list of all residents of HH before to turn to q 2-13)*	10. Occupation status	11. What is the size of salary, pension, scholarships or income from entrepreneurial activities?		12. Did (NAME) live in your household all the time during the last three months?	13. How many days was (name) absent from your household in the last three months?
		Employee..............1 Self-employed...........2 Employed pensioner ....3 Old age pension..........4 Disability pension........5 Pension on lost of supporter................6 Pension on special work condition..............7 Military service pension..8 Students................9 Employed students...10 Pupils..................11 Employed pupils.........12 Unemployed..........13 >12 No status ....99>12	Salary ....1 Income from entrepreneurial activities ....2 Pension ....3 Scholarship 4		Yes…1 -end of interview No…2	Number of days
			Amount (soms)	Code		
1						
2						
3						
4						
5						
6						
7						
8						
9						
10						
11						
12						

Interviewer! Enter the total number of household's residents _____
Out of them: the number of available household members _____
the number of absent household members _____
the number of outsiders _____

- 3 -

166

Confidentiality is guaranteed by the recipient of this information
Section II. EDUCATION OF THE HOUSEHOLD MEMBERS

[INTERVIEWER:] Use the list of names from the control card and make sure that numbers are correct.

From Q6 up to 10 ask children aged 6 – 18 only

No (from the control card)	Name	1. What is you education?	2. What field of science did (do) you study?	3. Do you study at the present period of time?	4. At what educational establishment do you study?	5. Indicate the type of educational establishment	6. In what class do you study?	7. For what reason you do not study?
		Doctor of science...1 Candidate of science...2 Higher education...3 Not-finished higher education...4 Secondary special...5 Secondary (general)...6—q.3 Incomplete secondary...7—q.3 Elementary...8—q.3 Incomplete elementary education...9—q.3 No education...10—q.3 Other...11—q.3 Age 0-6...98 —> q.3	Economics...1 Medicine...2 Technical sciences...3 Natural sciences...4 Humanitarian sciences...5 Pedagogy...6 Culture...7 Military school, academy...8 Agriculture...9 Other...10	Yes...1 No, 2=>7 (if age from 6 to 18 years old) ⇓ **Next section,** (if age elder then 19 years old)	General school...1 Higher educational institution...2—next section Secondary special...3—next section Courses...4 –next section	General school...1 Boarding school...2 Gymnasia...3 Lyceum...4 Special school for the children with disabilities...5 Other...6	Write the current class	Financial difficulties...1 Too expensive...2 Does not want to study...3 Too long distance to go...4 Irrelevant age...5 Health problem...6 Need to work...7 Dismissal...8 Other reasons...9
1								
2								
3								
4								
5								
6								
7								
8								
9								
10								
11								
12								

- 4 -

Talas Oblast

Confidentiality is guaranteed by the recipient of this information

Section III. MIGRATION

No (from the control card)	1. What country are you the citizen of? Interviewer! Enter name of country. Code	2. Was you born in this settlement? Yes...1=>8 No...2	3. In which country was you born? Interviewer! Enter name of country. In not in KR :q5	4. In which oblast of KR were you born? Refer to codes of oblasts below	5. Place of birth: Capital...1 Oblast or rayon center...2 city...3 urban village...4 village...5	6. At what age did you move from the place where you were born? number of complete years	7. Why did you leave the place where you were born? family curcumstances 1 due to the job (low wage)...2 job seeking...3 education...4 marriage...5 mil. service...6 threat...7 dangerous environment...8 ethnic conflict.9 other...10=>q-10	8. For the last 10 years, do you live in other settlement more than 3 months, apart from the current place of living? Yes...1 No....2 =>13	9. Where have you moved to your current place of living from? From the settlement where you was born, 99 In other cases, refer to the codes of oblasts below and state name of country	10. Place of your last living was... Capital...1 Oblast or rayon center...2 city...3 urban village...4 village...5	11. Why have you moved to current place of living? family curcumstances.1 due to the job (low wage)...2 job seeking...3 education...4 marriage...5 mil. service...6 threat...7 dangerous environment...8 ethnic conflict.9 other...10-q10	12. How long have you been living in this settlement since your last move? YEARS If less than year, state 0	13. Are you registered in this settlement? Yes...1 No....2
1													
2													
3													
4													
5													
6													
7													
8													
9													
10													
11													
12													

CODES OF OBLASTS OF KYRGYZSTAN

BISHKEK....41711, ISSYK-KUL....41701, JALAL-ABAD....41702, NARYN....41703, BATKEN....41704, OSH....41705, TALAS....41706, CHUI....41707, CHUI....41708

SECTION IV
DWELLING CONDITIONS

1. What type of dwelling do you have? ♦

A separate house 1
House part 2
Other type of dwelling 3
Temporarily dwelling 4
Apartment.............................5
Other uninhabited dwelling
used for residing 6

2. A form of ownership of your dwelling? ♦

State 1
Private 2
Rent ... ... 3⇒ q.4
Other 4

3. How did you received this dwelling? ♦

Received from state ... ...1⇒ q. 7
Bought from a private person ... ...2⇒ q. 7
Built it... ... ...3⇒ q. 7
Exchanged ... ... 4⇒ q. 7
Inherited ... ... 5⇒ q. 7
Other ... ... 6⇒ q. 7

4. Did you pay for rent? ♦

Yes 1
No ... ... 2⇒ q. 7

5. What amount in a month do you pay for rent? ♦

Amount (soms).................... [   ]
DN....... 9

6. Does the payment include the public utilities? ♦

Yes 1
No 2

7. What is the size of the dwelling that your family occupies? (with accuracy up to 1square meter) ♦

Square meters ... ... | _ | _ | _ |, | _ |

8. What is the size of the living space occupied by your family? (with accuracy up to 1square meter) ♦

Square meters ... ... | _ | _ | _ |, | _ |

9. How many living rooms does your family occupies? ♦

Number of rooms ... ... | _ | _ |

10. How long have you lived in this house? ♦

Write zero if less than one year
Complete years..... ....| | |

11. When was your house constructed? ♦

In _ | | | |
DN...9

[Interviewer! Ask Q. 13 if dwelling was constructed before 2000.]

12. Were there major repair works in your house? ♦

Yes..................................... 1
No....................................2 ⇒q.14
DN.................................9 ⇒q.14

13. When were capital repair works done in your house? ♦

Capital repair was done in _ | _ | _ | _ |
DN...9

14. From what construction materials are the walls of your dwelling mainly built? ♦

Bricks.................1
Concrete...............2
Adobe.................3
Wood..................4
Felt....................5
Clay....................6
Other................7 Specify_____

15. What is the main construction material of your roof?

Roofing slate...............1
Metal...........................2
Cane...........................3
Tile.....................4
Wood..................5
Abode..................6
Other................7 Specify_____

16. Availability of amenities ♦

No	Utilities	Yes - 1	No – 2=>next type
1	Heating		
A	Gas		
B	Electricity		
C	Coal		
D	Other fuel		
2	Telephone line fixed		

17. Was a power –meter installed? ♦

Yes..1
No...2=>q. 19
The electricity is absent ... ... 3=> q. 20

18. Is this meter used by your household only or jointly with other households? ♦

Only by one household 1
Jointly with other households........ 2

19. How often is power cut off in your household? ♦

Never disconnect 1
Some times in a year 2
Once a month 3
Once a week 4
Some times in a week 5
Every day 6

20. What water sources do you use? ♦

Water pipe...1
A well in the yard 2
Closed well...3
Water pipe in the yard..........................4
Public well, water pipe..........................5
A spring, river, lake, pond, irrigation ditch....6
Water is brought to us 7
Other 8

21. Where is the water source located? ♦

In house (flat)....................1
In yard............................2
Outside the dwelling or its
compound.........................3

22. How far is the water supply source from your dwelling? ♦

Less than 100 m.........................1
100-200 m...............................2
200-500 m...............................3
500-1000 m..............................4
More than 1000 m........................5
DN.....................................9

23. How often is water supply cut off in your household for last year? ♦

Never disconnect 1
Some times in a year 2
Once a month 3
Once a week 4
Some times in a week 5
Every day 6

24. For how many months for last winter was your house heated? ♦

Number of months..........................[   ]

25. Where do your family members have a bath? ♦

A bathroom with a shower....................1
Shower.......................................2
Private bath sauna 3
Public sauna, bath............................4
Other...5

26. What type of toilet do you have in your household? ♦

Toilet with individual sewage system......1
Pit latrine 2
Other type of toilet.........................3
No toilet....................................4=> q.29

27. Where is the toilet situated? ♦

In the dwelling 1
Outside............................2
In the street..........................3

28. Is this toilet used by your household only or jointly with another one? ♦

By 1 household only 1
Jointly with another one 2

29. How does your household get rid of garbage? ♦

Refuse chute................................1
Lorry-container collection.................2
Throwing to garbage piles.................3
Burning.....................................4
Burying.....................................5

30. What do you use for cooking? ♦

Primus stove...............................1
Gas stove with balloons 2
Floor standing electric flier............3
Electric flier..............................4
Stove......................................5
Other......................................6

Talas Oblast   *Confidentiality is guaranteed by the recipient of this information*

31. How much does it take you to reach the nearest public bus stop?♦
Less than 5 minutes.............................1
5-15 minutes......................................2
15-30 minutes....................................3
30-60 minutes....................................4
More than 1 hour...............................5

Interviewer! Do not ask Q 32-33 if respondents rent the dwelling.

32. Evaluate, for how much could you sell your dwelling?♦
Sum (in soms)............................[   ]
DN..9

33 For how much per month could you lease your dwelling?
Non-furnished ... [   ]
Furnished ....[   ]
DN...9

34. Do you have an additional dwelling?
Yes..1
No...2 => next
section

35. What dwelling is this?
A separate flat....................................1
House or half a house in rural area 2
House or half a house in urban area.....3
Dacha..4
Other..5

SECTION V

AVAILABILITY OF DURABLES

1. How many durables do you have in your household? (in working condition)♦

No	Durable	Quantity
1	A	2
1	Radio-receiving set	
2	Musical centre	
3	White&black TV-set	
4	Colour TV-Set	
5	Video	
6	Tape-recorder	
7	Video-player	
8	Video-camera	
9	Musical centre	
10	Photo camera	
11	Personal computer	
12	Printer	
13	Ordinary washing machine	
14	Automatic washing machine	
15	Vacuum cleaner	
16	Bicycle	
17	Refrigerator ordinary	
18	Refrigerator with 2-3 camera	
19	Freezer	
20	Electrowave	
21	Microwave	
22	Electric heater	
23	Table	
24	Chairs	
25	Bed	

26	Couch	
27	Arm-chair	
28	Arm-chair-bed	
29	Wool Carpet	
30	Synthetic carpet	
31	Rug, floor cover	
32	Cell-phone	
33	Wardrobe	
34	Book-case	
35	Side-board, wall furniture	
36	Book shelf	
37	Corner furniture	
38	Shelves	

Auto machinery

No	Type	Quantity	Production date	Date of purchase	Market price, soms
1	A	2	3	4	5
1	Car				
2	Lorry				
3	Minibus				
4	Motorcycle, motor-scooter, motorbike				
5	Motor-bicycle				

- 8 -

171

Talas Oblast   Confidentiality is guaranteed by the recipient of this information

31. How much does it take you to reach the nearest public bus stop? ♦

Less than 5 minutes............................1
5-15 minutes......................................2
15-30 minutes....................................3
30-60 minutes....................................4
More than 1 hour...............................5

Interviewer! Do not ask Q 32-33 if respondents rent the dwelling.

32. Evaluate, for how much could you sell your dwelling? ♦

Sum (in soms)...............................[   ]
DN..9

33 For how much per month could you lease your dwelling?

Non- furnished ... [   ]
Furnished ....[   ]
DN...9

34. Do you have an additional dwelling?

Yes...1
No...2 => next section

35. What dwelling is this?

A separate flat.....................................1
House or half a house in rural area ...2
House or half a house in urban area....3
Dacha..4
Other...5

SECTION V
ﻝﻯﻝﻯﻝﻯﻝﻯﻝﻯﻝﻯﻝﻯﻝﻯﻝﻯﻝﻯ

AVAILABILITY OF DURABLES

1. How many durables do you have in your household? (in working condition) ♦

No	Durable	Quantity
1	A	2
1	Radio-receiving set	
2	Musical centre	
3	White&black TV-set	
4	Colour TV-Set	
5	Video	
6	Tape-recorder	
7	Video-player	
8	Video-camera	
9	Musical centre	
10	Photo camera	
11	Personal computer	
12	Printer	
13	Ordinary washing machine	
14	Automatic washing machine	
15	Vacuum cleaner	
16	Bicycle	
17	Refrigerator ordinary	
18	Refrigerator with 2-3 camera	
19	Freezer	
20	Electrowave	
21	Microwave	
22	Electric heater	
23	Table	
24	Chairs	
25	Bed	

26	Couch	
27	Arm-chair	
28	Arm-chair-bed	
29	Wool Carpet	
30	Synthetic carpet	
31	Rug, floor cover	
32	Cell-phone	
33	Wardrobe	
34	Book-case	
35	Side-board, wall furniture	
36	Book shelf	
37	Corner furniture	
38	Shelves	

Auto machinery

No	Type	Quantity	Production date	Date of purchase	Market price, soms
1	A	2	3	4	5
1	Car				
2	Lorry				
3	Minibus				
4	Motorcycle, motor-scooter, motorbike				
5	Motor-bicycle				

- 8 -

172

Talas Oblast

Section VI. Agricultural Activity
6.1. Land

1. Do you have a land plot? Yes ☐ ... No ☐   (1 / 2)

2. Did you use your land for agricultural activity? Yes ☐ ... No ☐   (1 / 2)

3. If land is not used explain reason?
- no money — 1
- no labour — 2
- do not want — 3
- unprofitable — 4
- I do not wish to answer — 5
- Other reason — 6

Land use

I would like to know what kinds of land resources are in your ownership?

		4		5	6	7	8		9	10		11
		Size of land plot		Do you have land ownership documents?	What is estimated cost of land?	What a source of irrigation do you use?		Do you lease your land to rent?	Specify the land in rents quantity		How much did you receive as a rent? (Including cost of products and/or services)
		One hundred m2 - 1 Hectares - 2		Yes - 1 No - 2	soms	Canal – 1 Water reservoir – 2 Pump – 3 River, water –ditch - 4 Rain - 5 Other - 6		Yes - 1 No - 2	One hundred m2 - 1 Hectares - 2		soms
Code	Types of land resources	Quantity	Code			Code			Quantity	Code	
1	Arable land irrigated										
2	Non irrigated land										
3	Haymakings										
4	Gardens										
5	Personal plot										
6	Other										

13. Do you rent a land? Yes ☐ No ☐   (1 — ; 2 go to the following section)

14. What kinds of land do you rent?

		15		16	17	18	19	20	21	22	23
		Size of rented land plot			Do you have land rent documents?	How much do you pay for land including tax as a rent? (Including cost of products and/or services)		What a source of irrigation do you use?	Do you want to purchase rented land plot?	Specify a sum you would like to buy a site?
		One hundred m2 - 1 Hectares - 2			Yes - 1 No - 2	soms		Canal – 1, Water reservoir – 2, Pump – 3, River, water –ditch - 4, Rain – 5, Other - 6	Yes - 1 No - 2	soms
Code	Types of land resources	Quantity	Code							
1										
2										
3										

- 1 -

173

Talas Oblast

Confidentiality is guaranteed by the recipient of this information

6.2. Characteristics of the land resources

1 Do you want to increase your site?
Yes
No
I do not know

3 What the form of ownership is better from your opinion?
Rent
Private ownership

2 If yes, in explain how:
buy the neighbour site
buy any accessible site
rent a site
develop the infertile sites
other ways respectfy)

4 How much do you ready to invest in the land in the following 5 years?

5 Did you bought the land within the last three years?
Yes
No
soms

5a Specify the cost and the size of a site

Type of a land
The size
Cost

Quality of the land

6 How are you estimate the quality of your land plots?

	Land share	Personal plot	Other
Excellent	4		
Good	3		
Average	2		
Low quality	1		
I do not know	0		

7 What from the following problems are relevant for your land?:

	Yes	No	I do not know	8
Fertility				
Saltiness				
Marshiness				
Rockiness				
Water supply				
Access to a site				

8 What is the distance from your site to the house?

		Land share	Other
less than 1 km	1		
from 1 to 5 km	2		
from 5 to 10 km	3		
more then 10 km	4		

9 Could you specify the problems caused by the distance toy to your site?
Yes
No
I do not know

10 How do you estimate a condition of irrigation system?
Excellent 4
Good 3
Average 2
Low quality 1
I do not know 0

How many times do you water during a season ....
Arable land
Personal plot

11 Do you have a problems with irrigation in a season?
Yes
No

12 How much do you pay for irrigation, a som per hectar? soms

13

14 Dou you implement any measures to improve the quality of land?
Yes
No

15 If yes, specify? ....
mineral fertilizers
organic fertilizers
exhaust, a chalk
drainage works
soil alignment
fence installation

- 2 -

Talas Oblast

6.3. Cultivation and use of agricultural crops

Has your household been involved in any crop production activities?

Yes 1
No 2 next section

1. Crop cultivated at the main plot

Specify harvest for the current year, if not available, mention harvest for the last year

HH- Household

Code	Name	Area One hundred m2 - 1 Hectares - 2		Harvest	How many kg did your HH* sell?	How much money did your HH* receive from sale?	How many kg did your HH* keep for seeding?	How many kg did your HH* lose to insects, rodents, fire or spoilage?	How many kg did your HH* consume?	How many kg did your HH* process	Rest of the product
		Quantity	Code	kg	kg	som	kg	kg	kg	kg	kg
1											
2											
3											
4											

How many kg did your HH* from the rest products...

Code	Name	Transferred as a gift	Exchanged	Give in credit	Other transfer specify please
		kg	kg	kg	kg
1					
2					
3					
4					

2. Crops cultivated at personal land plot or otehr additional land plot

Code	Name	Harvest	Consumed in summer period	Sold	Processed	Storaged
		kg	kg	kg	kg	kg
1						
2						
3						
4						
5						
6						
7						
8						
9						

- 3 -

175

Talas Oblast

Confidentiality is guaranteed by the recipient of this information

6.4. Land processing expenses

1. Does your HH(household) buys for agricultural activity....
(enter supplier code)

a Seeds?
b Mineral fertilizer/ pesticides ?
c Organic manure?
d Packaging material?

Supplier Code

Market	1 Neighbor/friend/relatives
Specialized supplier	2 Other source (Specify)
3 Do not buy	5
	4

2. Dis your HH buy inputs in credit?
Yes
No

4

3. What percentage of these inputs were bought in credit?

a Seeds?
b Mineral fertilizer/ pesticides ?
c Organic manure?
d Packaging material?

9

Code	Name	How much did your HH spend on seeds?	How many kilograms of mineral fertilizers did your HH purchase?	How much did your HH spend on mineral fertilizers?	How many kilograms of herbicides and pesticides did your HH purchase?	How much did your HH spend on herbicides and pesticides?	How many kilograms of organic manure did your HH purchase?	How much did your HH spend on organic manure?	How much did your HH spend on packaging materials?	How much did your HH spend on transportation?	How much did your HH spend on storage?
		If zero, write 0	*If zero -> q. 7*		*If zero -> q. 9*		*If zero -> q. 11*				
		som	kg	som	kg	som	kg	som	som	som	som
1											
2											
3											
4											
5											

14. How much did your HH spend on the following works?
IF NOTHING WAS SPENT WRITE ZERO

Mechanized works

		Rate per 1 ha		Land plot	Amount	already paid	Rest of the debt
		som	fuel (l)	ha	som		
a	Clearing land						
b	Plowing						
c	Seeding						
d	Harvesting						
e	Others						

Manual operations IF NOTHING WAS SPENT WRITE ZERO

	Rate per day, som	Amount of man-days	Total sum, som
a	Payment for hired workers for harvesting		

16. Harvesting by combine IF NOTHING WAS SPENT WRITE ZERO

		Rate per ha	Haricot beans, kg	Fuel
a	Rate per ha			
b	Total paid			
c	Paid for works			
d	Rest of the debt			

only for haricot beans

17. How much did your HH spend on other expenses?
IF NOTHING WAS SPENT WRITE ZERO

		som
a	Land Tax	
b	Social Fund	
c	Other taxes	
d	Other expenses	

- 4 -

176

Confidentiality is guaranteed by the recipient of this information

6.5. Food products produced **within the HH**

1. Did your HH produce any food products from the crops grown or bought by the HH?

Yes   1

No   2 -> next section

2. Which members of your HH helped in the production home produced food products?
If more then 3 members were involved, put codes of those who helped the most.

Code [＿＿＿]   Code [＿＿＿]   Code [＿＿＿]

Code	Name	3 Did your HH produce food from grown or purchased products? Yes - 1 No - 2 -> Next product	4 What quantity did your HH produce? Quantity	5 What quantity of the product were consumed by your HH? If zero -> q. 7 Quantity	6 How many months did your HH consume the product? Number of months	7 What quantity of the product was given away as gift or help to other peoples? If zero write 0 Quantity	8 What quantity of the product did your HH store? If zero write 0 Quantity	9 What quantity of the product did your HH sell? If zero -> q. 11 Quantity	10 How much money did your HH receive from the sell of the food? som	11 How much money did your HH spend on producing the product? som
1	Jam. 1									
2	Canned vegetables. 1									
3	Spiced vegetable pasta. 1									
4	Wheat flour. kg									
5	Juices. 1									
6	Compote. 1									
7	Dried fruits. kg									
8										
9										
10										
11										
12										
13										

Confidentiality is guaranteed by the recipient of this information

6.6. Livestock

1. Did your HH involved in livestock breeding?
   Yes 1
   No .... 2 => next section

2. Which members of your HH involved in raising animals?
   If more then 3 members were involved, put codes of those who helped the most.
   Code [   ] [   ] [   ] Code

N	A	3. How many heads of livestock did your HH have 12months ago? (Quantity)	4. How many heads of livestock were sold? (Quantity)	5. How much money did your HH receive from sale of the animal? (Som)	6. How many heads of livestock were purchased? (Quantity)	7. How much money did your HH pay for purchased animal? (Som)	8. How many animals were born and received as a gift? (Quantity)	9. How many animals did your HH slaughter? (Quantity)	10. How many animals were lost from diseases, stolen, or died? (Quantity)	11. How many animals did your HH given away for free (presented as a gift)? (Quantity)	12. How many heads of livestock did your HH have at present? (Quantity)	13. How much could you receive from sale of animals by current prices? (Quantity)
1	Cows											
2	Heifers over a year											
3	Bulls and bullocks over a year											
4	Calves younger than one year											
5	Pigs over a 9 months											
6	Piglets from 4 to 9 months											
7	Piglets younger than 4 months											
8	Ewes over one year											
9	Sheep younger than one year											
10	Sheep younger than one year											
11	Goats over a year											
12	Horses											
13	Poultry, young and old											
14	Rabbits											
15	Donkeys											
16	Other types of animal											

- 6 -

Talas Oblast

6.7. Animal products

1. Did your HH produced any animal products?
   Yes   1
   No   2 -> next section
2. Which member of your HH were involved in producing animal products?
   If more then 3 members were involved, put codes of those who helped the most.
   Code [Code]

N n/n	A	What quantity of the product were produced by your HH in the past 12 months?	How many months did your HH consume the product?	What quantity of the product were consumed by your HH in the past 12 months?	What quantity of the product were given away for free (presented) by your HH in the past 12 months?	What quantity of the product did your HH store?	What quantity of the product did your HH sell in the past 12 months?	How much money did your HH receive from the sell of the food?	How much money did your HH spend on producing the product?
		3	4	5	6	7	8	9	10
		Quantity	Months	Quantity	Quantity	Quantity	Quantity	som	som
1	Mutton, lamb								
2	Beef								
3	Poultry								
4	Eggs								
5	Milk								
6	Sub-products								
7	Wool, fur								
8	Skins								
9	Honey								
10	Fish								
11	Other products								
12									

- 7 -

179

6.8. Livestock expenses

1. Veterinary costs

№	Expenses	Quantity, heads	Rate per 1 head, som	Total
1	Sheep			
2	Cattle			
3	Horses			
4	Poultry			
5	Other (specify)			

2. Livestock herding expenses

№	Expenses	Quantity, heads	Rate per 1 head per months, som	Quantity of months	Total
1	Village ranges herding				
а	Sheep				
6	Cattle				
в	Horses				
2	Remote ranges herding				
а	Sheep				
6	Cattle				
в	Horses				

3. Fodder purchasing expenses

№		Quantity, kg	Price per 1 kg, som	Total
1	Grain			
2	Hay			
3	Salt			
4				
5				
6				
7				
8				

4. Other expenses

№	Describe	Total, som
1		
2		
3		
4		
5		
6		
7		
8		

Confidentiality is guaranteed by the recipient of this information

6.9. Do you have any agricultural equipment in your household (including horses)? ◆
Yes 1
No ....2=> q.12

10. What kind of agricultural equipment in your household, is it rented or your property?

N n/n	Name	Quantity	Private		Rented	
			Used by 1 HH - 1	Used by several HH - 2	Used by 1 HH - 1	Used by several HH - 2
	A		2	3		4
1	Working horses					
2	Donkey					
3	A cart					
4	Water-pump					
5	Seeder					
6	Tractor					
7	Motor-block					
8	Mini-tractor					
9	Mower					
10	Combine					
11	Plough					
12	Other (specify)					
13						

- 8 -

Talas Oblast

Confidentiality is guaranteed by the recipient of this information

Section 7. Food expenditures

1. Were any HH member absent from the household within last 14 days?

Yes   -1

No   (2) →   q. 4   Code (_____)

2. Specify who was absent more then 2 days within last 14 days?

Code (____) (____) (____) (____)

3. How many days were absent specified HH members within last 14 days?

Days (____) (____) (____) (____)

11. Did your HH purchase products in credit?   Yes -1

                                             No   -2

12. Specify hou much did your HH purchase   (____) som

food in credit for the last 14 days?

13. Did your HH exchange the food products on Yes -1

barter to the other goods or products?   No   -2

14. Specify the sum of the barter exchanges (____) som

within the last 14 days?

№	Product	3 Did your HH purchase the product within the past 12 months? Yes -1 No- 2 Code	4 How many months within the past 12 months your HH purchase the product? Number of months	5 How often did you buy this product 1 - daily, 2 - once a week 3 - Once a month 4 - every 3 month 5 - once a year Code	6 How much did your HH normally buy? gram - 1 kg - 2 liter - 3 item - 4 Quantity Code	7 How much money does your HH normally spend on product. when purchase it? som	8 Where did you buy the product? Bazaar - 1 Shop - 2 Private market - 3 Private person - 4 Other - 5 Code
1	White bread						
2	Lepzoshka (national bread)						
3	Wheat Flour						
4	Rice						
5	Peas						
6	Pasta, noodles						
7	Potatoes						
8	Carrots						
9	Onion						
10	Tomatoes						
11	Cucumbers						
12	Garlic						
13	Pepper						
14	Hot pepper						
15	Cabbage						
16	Other vegetables						

- 9 -

181

Talas Oblast

Confidentiality is guaranteed by the recipient of this information

№	Product	3 Did your HH purchase the product within the pat 12 months? Yes -1 No - 2 Code	4 How many months within the past 12 months your HH purchase the product? Number of months	5 How often did you buy this product 1 - daily, 2 - once a week, 3 - Once a month, 4 - every 3 month 5 - once a year Code	6 How much did your HH normally buy? gram - 1 kg - 2 liter - 3 item - 4 Quantity	Code	7 How much money does your HH normally spend on product, when purchase it? som	8 Where did you buy the product? Bazaar - 1 Shop - 2 Private market - 3 Private person - 4 Other - 5 Code
17	Apples							
18	Pears							
19	Tropical fruits							
20	Other local fruits							
21	Dried fruits							
22	Water-melon							
23	Melon							
24	Jam							
25	Tomato-paste							
26	Baby food							
27	Canned food vegetable							
28	Honey							
29	Nuts and sunflower seeds							
30	Mutton							
31	Beef							
32	Horse-flesh							
33	Cows offal							
34	Horse byproducts							
35	Sausages							
36	Hens and other fowl							
37	Fresh fish							
38	Canned fish							
39	Canned meat							
40	Eggs							
41	Milk							
42	Kefir							
43	Airan							
44	Yogurt							

- 10 -

182

Confidentiality is guaranteed by the recipient of this information

№	Product	3 Did your HH purchase the product within the pat 12 months? Yes -1 No- 2 Code	4 How many months within the past 12 months your HH purchase the product? Number of months	5 How often did you buy this product? 1 - daily, 2 - once a week 3 - Once a month 4- every 3 month 5 - once a year Code	6 How much did your HH normally buy? gram - 1 kg - 2 liter - 3 item - 4 Quantity	Code	7 How much money does your HH normally spend on product, when purchase it? som.	8 Where did you buy the product? Bazaar - 1 Shop - 2 Private market - 3 Private person - 4 Other - 5 Code
45	Butter							
46	Cheese							
47	Margarine							
48	Mayonnaise							
49	Vegetable oil							
50	Animal fat							
51	Tea							
52	Coffee							
53	Alcohol drinks							
54	Maxim, Bozo							
55	Juices							
56	Carbonated beverages							
57	Sugar							
58	Salt							
59	Pepper and other spices							
60	Confectionaries							
61	Cigarettes							

183

SECTION 8. NON-FOOD EXPENDITURES AND SERVICES

1. Expenditure on purchase of non-food items and services

Lets talk about all expenditure categories

#	Type of expenses	Last three months (Som)	#	Type of expenses	Last three months (Som)
1	A	3	1	A	3
1	Clothes, underwear		11	Jewellery	
2	Hats		12	Vehicle related costs	
3	Head-dresses (scarf, kerchief).		13	Big electric appliances	
4	Gloves mittens		14	Kitchen dishes	
5	Hosiery		15	Furniture	
6	Foot wear		16	Household textile items	
7	Fabrics		17	Agricultural equipment	
8	Accessories and clothes decorations		18	Construction materials and sanitary appliances	
9	Leather accessories		19	Other non-food items	
10	TV, radios and other entertainment-equipment		20	Total	

2. UTILITY EXPENDITURES

1. Do you live in a private house? Yes..1

   No...2⇒ 3

2. Did you have expenditures related with purchase of fuel for the last three months?

#	Type of fuel	Units: 1 cub.m. 2 – kg 3 – c 4 - t	What quantity did you buy for the last three months?	Cost of purchases for the last three months (soms)	How much did you use for heating?	How much did you storage yourself?	How much did you receive as a gift?
1	A	2	3	4	5	6	7
1	Wood						
2	Peat						
3	Coal						
4	Mazut						
5	Other						

- 12 -

184

2. Payment for utility services and electricity taking into account benefits and without taking it into account (answer the families that live private houses and apartments)

#	Type of service	Do you use the listed bellow services? Yes.....1 No......2⇒ Next service	How much did you pay? (soms)	
			Summer time	Winter time
1	A	2	3	4
1	Electricity			
2	Cold water and sewage			
3	For refuse collection			
4	Other utility services			

3. HEALTH CARE EXPENDITURES

1. Did anyone from you household pay for any medical appliances (glasses, hearing device, crutches, etc.), medicines, herbs or visits to healthcare institutions or hospitalization in the last three months?

Yes..1

No...2 section 4

DN...9 section 4

2 What have you spent this money for? (Som) (Payment could in cash and in-kind)

#	Type of expenditure	Code of the member family	Last three months (Som)
1	A	1	3
1	Medical appliances (crutches, wheelchair, warmers, contraception, contact lenses, etc)		
2	Medicine		
3	Consultation fees		
4	Visit to a dentist		
5	Laboratory tests, ultrasound, diagnostic examination, physiotherapy, massage, injections, etc		
6	Official co-payment for admission to doctors		
7	For food		
8	Medicine		
9	Informal payment to doctors and medical staff (including gifts)		
10	Lab tests, ultrasound, x-ray, physiotherapy, etc		

- 13 -

185

11	Bandages, syringes, injections, bed sheets, and other materials		
12	Other expenditures (describe in detail)		

4. TRANSPORTATION EXPENDITURES

I would like to talk about your transportation expenditures.

1. Did your family use the services of the suburban or long distance transport? (except, municipal transport and transportation expenses relating to education?)

Yes...1

No..2⇒ section5

DN...99⇒ section 5

2. What type of transport did your household use and how much did you pay for these services? ⇓(Som)

№ /п	Transport type		Last three months (Som)	Family Member Code
1		A	5	
1	Bus transport	Within the Republic		
2		Outside the Republic		
3	Taxi	Within the Republic		
4		Outside the Republic		
5	Railway	Within the Republic		
6		Outside the Republic		
7	Airway	Within the Republic		
8		Outside the Republic		

5. EDUCATIONAL EXPENDITURES

1. Did you have any expenses on childcare and on pre-school facilities?

Yes..1

No...2⇒ 3

2. For what did you pay? (Som)

Code	Service type	Last 12 months (Som)
1	A	4
1	Payment for pre-school facilities	
2	Purchase of pre-school accessories	
3	Nanny's services	
4	Unofficial expenses (gifts)	

- 14 -

5	Other expenditures (describe in detail)	

3. Did you have any expenses for education for last 12 monthes?

Yes..1

No..2⇒ section 6

DN..9⇒ section 6

4. How much did you spend for education in total for the last twelve months?

Type of educational institution			At the elementary school	In secondary school	Vocational school	Higher educational establishment	Courses
Do you use the following services?	Yes— 1	No—2					
Payment for school							
Text-books, school stationary							
School uniform and sports wear							
Use of library							
Coaching and additional lessons							
Transportation expenses							
School maintenance (repair works)							
Unofficial expenses							
Other expenses							
Total expenses							

6. OTHER EXPENDITURES

1. Did your household have any of the following expenses? ⇓(Som)

Code	Type of expenses	Last twelve months (Som)
1	A	4
1	Sanatorium- resort Tourist- exhibition services	
2	Various insurance payments	
3	Fees from owners of transportation means	
4	Legal services	
5	Weddings, funerals and other ritual services	
6	Gifts on celebrations, weddings, funerals to relatives, colleagues, friends etc.	
7	Communication cervices(including stationary telephone services) internet, radio (last three month average expenses)	
8	Mobile communication (last three month average expenses)	

9	Other services (in detail)	

2. Did your family pay alimonies?

Yes...1

No...2⇒ 4

3. What amount of money that your household paid for alimonies? ⇓

	Last twelve months (Som)
Sum (soms)	

4. Does your family help relatives and friends, who do not live with you?

Yes...1

No...2⇒ section 7

5. What type of help was it? If it was money, what sum did you receive? If it was in-kind help, could you try to assess its value? ⇓

Code	Type of help	In the last month	Two months ago	Three months ago
1	A	2	3	4
1	Money			
2	Food products produced by yourself			
3	Food products bought			
4	Clothes, foot wear			
5	Medicine			
6	Other (indicate in detail)			

- 16 -

188

9. Eating out, other sources of food

1. Has any member of your HH eat or purchase drinks outside of the home

Yes 1
No 2 -> q. 2

If more then 3 members were involved, put codes of those who is active the most.

№	Code of the member ( )		
	Average quantity per month	Average price, som	Total sum, som
1	Breakfasts		
2	Lunches		
3	Dinner		
4			

№	Code of the member ( )		
	Average quantity per month	Average price, som	Total sum, som
1	Breakfasts		
2	Lunches		
3	Dinner		
4			

№	Code of the member ( )		
	Average quantity per month	Average price, som	Total sum, som
1	Breakfasts		
2	Lunches		
3	Dinner		
4			

2. Did your HH receive food for free from friends, relatives or humanitarian organisations for the last 12 months?

Yes 1
No 2 -> Next section

3. How many times did you receive food for the last 12 months?

Date	Source of help	Product name	Quantity

4. Can you please estimate the cost of food received by your HH for free?   som

5. Did you expect the same food support within the next 12 months?

Yes 1
No 2 -> Next section

6. Did you expect that the support will be the same as in the last 12 months?

Yes 1
No 2
Do not know 3 -> Next section

- 17 -

189

Section 10. Loans and savings

1. Did your HH make loans to friends, relatives or business partners?
    Yes..1
    No...2==> q.3
    DK..9==> q.3

2. 30. How much did these persons still owe to your HH?

    Total amount (_____) som

3. Have you or members of your HH purchase
    any goods or services in credit within last 12 months?
    Yes..1
    No...2==> Section10.1.

4. What is the value of the goods or services purchased by your HH in credit?

    Total amount (_____) som

5. How much do the members of your HH still owe
    for the goods purchased in credit?

    Total amount (_____) som

Section10.1. Savings

1. Does any member of your HH have any savings in cash or
    in account in the bank?
    Yes   1
    No   2

2. If you have savings in the bank account, please specify the value of deposit?
    1 > less then 50 000 som
    2 from 50 000 to 100 000 som
    3 from 100 000 to 500 000 som
    4 more then 500 000 som
    9 DN

3. Could you specify the starting date of the deposit?
    _____

4. Specify please, the interest rate
    _____ %

5. Do you think that bank condition is satisfactory?
    Yes   1
    No   2

Confidentiality is guaranteed by the recipient of this information

10.2. Loans received by the members of the household

	Is the lender of your loan? Relative...1 Private person...2 State bank...3 Commercial bank...4 Microcredit organisation...5 Other...6	Value of the loan	When did you receive a loan?		Purpose of loan Agricultural operations...1 Purchase of land...2 Purchase of amenities...3 Other business...4 Construction or purchase of immovable property...5 Special events...6 Education...7 Other...8	Interest rate	Did your HH give a collateral for received loan? Yes......1 No......2 ->question 8	What was used as a collateral? Dwelling.........1 Land.......2 Jewelry...3 Personal items...4 Other...5
	1	2	3		4	5	6	7
		Som	month	year		%		
A	Code				Code			Code
1ST LOAN								
2ND LOAN								
3RD LOAN								
4TH LOAN								

	Did you make repayment of loan to the lender? Yes......1 No......2 ->q.11	How much did your pay and how often? weekly...1 monthly...2 quarterly...3 half year...4 year once...6		Was payment of this loan were repaid weekly? Yes......1 No......2 ->q.11	Rest of the debt by loan?	Date of the final repayment ecaulf as fixed date, write zeros	
	8	9		10	11	12	
		som	periodicity		som	month	year
A							
1ST LOAN							
2ND LOAN							
3RD LOAN							
4TH LOAN							

Section 11. Non - agricultural activity

1. Did any member of your HH work as self-employed entrepreneur or hired labour

Yes ... 1 Code of the HH's memmber

No ... 2

2. Describe the type opf activity Code of the HH's memmber
   a trade by food products
   b trade by non- fuud stuff
   c transport services
   d processing of agr. product
   e hired seasonal labour in agriculture
   f other (specify)

3. Did you work independently or work as hired labour?
   Code of the HH's memmber
   a Work independently
   6 Hired labour -> q. 8

4. Specify the amount of the income received from your entrepreneur
   activity for the last 12 months?

Number of months	
Monthly gross income (without expenses)	
Total gross income	

5. Did you hire additionla labour?
   Yes ... 1
   No ... 2 -> q. 7

6. Specify please average monthly salary
   and amount of the months
   hired labour for the last 12 months?

Number, persons	
Average salary	
Number of months	
Total expenses on salary	

7. Describe other expenses related to the business activity

	Monthly sum, som	Number of months	Total sum per year, som
Patent			
Renting equipment			
Renting working space, office, trading place			
Expenses on th equipment repair			
Taxes (Social Fund)			
Other expenses			
Total sum of additional expenses			

8. If you are hired worker, please specify
   the salary and number of month you are
   working within last 12 months

	Code
Number of months	
Average salary, som	
Total salary per year, som	

11.2. Labor Migration

9. Does you family labor migrants?
   Yes [ ] No [ ]

10. How many years / months ago labour migrant is leave your family?
    Year [ ] Month [ ]

11. What is the country where labour migrant reside, and the sector working?
    country [ ] Sector [ ]

12. Did you receive any transfers in a last 6 months from a labor migrant?
    Yes [ ] No [ ]

13. How much was the sum of the transfer?
    Sum [ ]

- 20 -

192

B   Supporting Letter

КЫРГЫЗ РЕСПУБЛИКАСЫНЫН ӨКМӨТҮНҮН АППАРАТЫ	АППАРАТ ПРАВИТЕЛЬСТВА КЫРГЫЗСКОЙ РЕСПУБЛИКИ
720003, Бишкек ш., Абдумомунов көч., 207	720003, г. Бишкек, ул. Абдумомунова, 207

201__-ж.г. «_4_» _05_____   № Н-1321

Главе государственной
администрации Таласской
области – губернатору
Курманалиевой К.А.

Уважаемая Койсун Аблабековна!

Настоящим уведомляем, что в рамках исследовательского проекта Университета Юстус-Либих, Гиссен, Германия проводятся социально-экономические обследования уровня жизни сельского населения в Кыргызской Республике. В настоящее время в реализации данного проекта в Таласскую область направляется Тилекеев Канат Асанович. Результатом данного проекта будет выработка рекомендаций и предложений для совершенствования государственной политики в области социальной защиты населения в отдаленных сельских регионах Кыргызской Республики. Для осуществления данной задачи просим ему оказать необходимое содействие в поиске и получении необходимой информации, а также организацию встреч с заинтересованными организациями по Таласской области.

С уважением,
Заведующий отделом
организационно-инспекторской работы
и территориального управления   С.Уметалиев

194

Office of the Government of the Kyrgyz Republic
Abdymomunova Str., 207
720003, Bishkek
Kyrgyz Republic

April 4, 2011   No.11-1321

To: Kurmanalieva K.A.,Governor of Talas Oblast State Administration
Dear Koisun Ablabekovna!

With this letter we would like to inform you that within the framework of a research project
of Justus Liebig University Giessen, Germany, a survey of the social-economic standards of
living of the Republic's rural population is to be conducted. Within this project's framework,
Talas Oblast has been allocated to Kanat Tilekevev. The result of this project will help to
develop recommendations and proposals to improve the public policy in the field of social
protection of the population in remote rural areas of the Kyrgyz Republic. To support the
successful project work, we would ask you to provide him the necessary assistance in locating
and obtaining the necessary information, as well as in organizing meetings with interested
organizations in Talas Oblast.

Sincerely,
S. Umetaliev
Head of the Department of Organization
and Territorial Management

C   Maps of the Sampled Villages

The maps of the 30 villages covered during the field study is presented in this appendix. Certain adjustments of the maps were made to provide the privacy of the interviewed households. Names of the villages and streets were deleted from the maps. Visited households are marked by black circles on the area covered by the respective village. The maps catalogue, including the names of the villages and the addresses of the households, are stored at the Econometrics and Statistics Chair of the Economics and Business Faculty of Giessen University:

Justus-Liebig-Universität Gießen
Fachbereich Wirtschaftswissenschaften
Professur für Statistik und Ökonometrie
Licher Straße, 64
D-35394 Gießen

Village 1

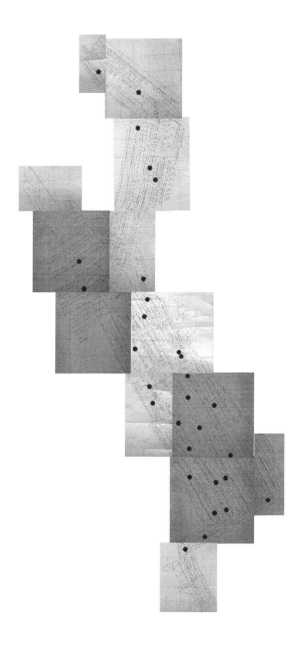

Village 2

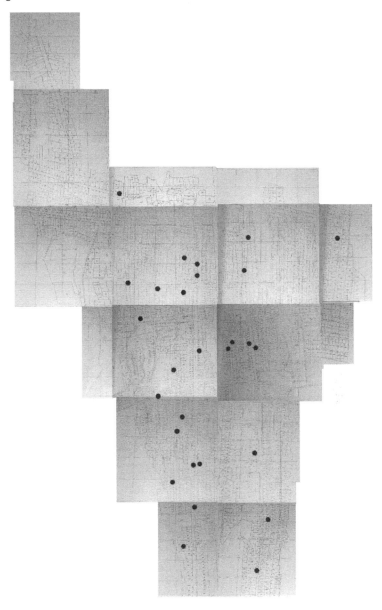

Village 3

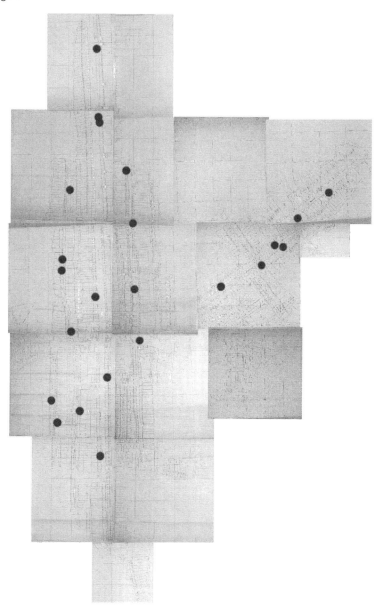

Village 4

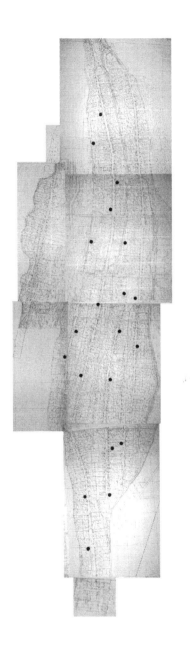

Village 5

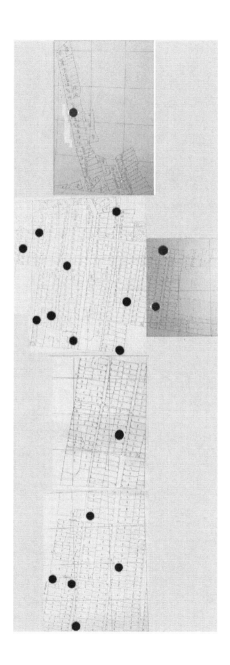

Village 6

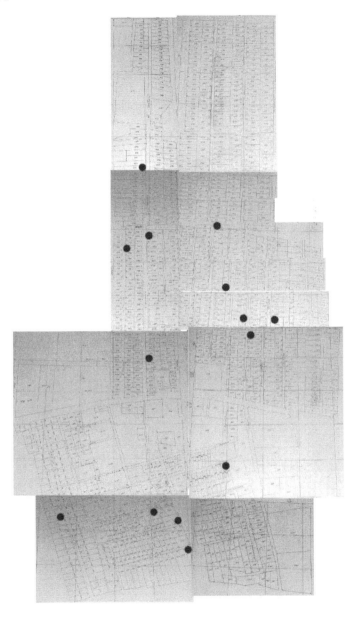

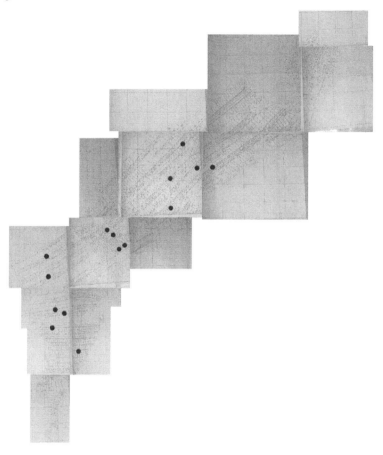

Village 8

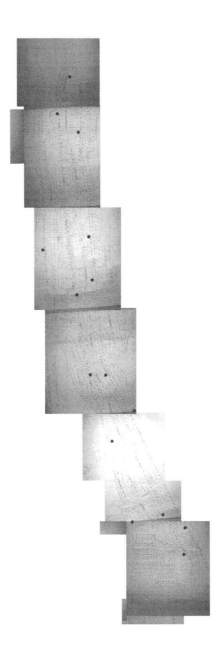

Village 9

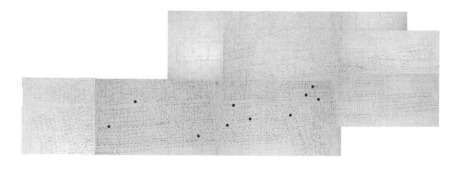

Village 10

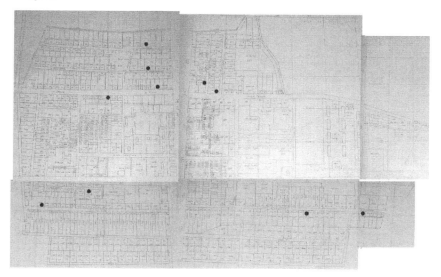

Village 11

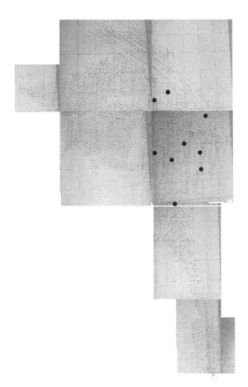

Village 12

Village 13

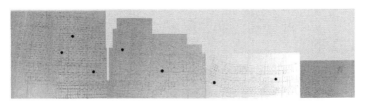

Village 14

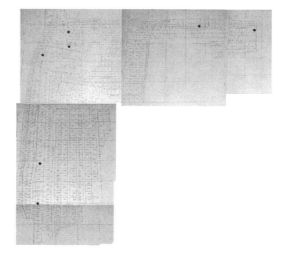

Village 15

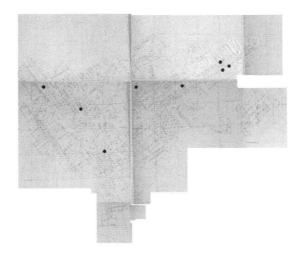

Village 16

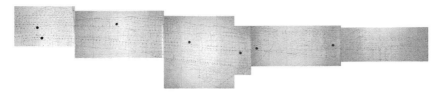

Village 17

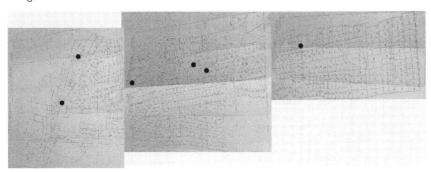

Village 18

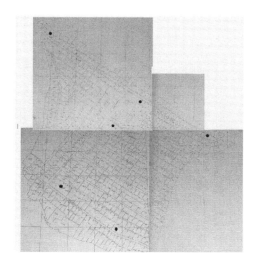

Village 19

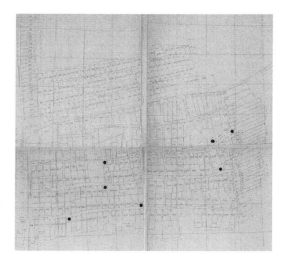

Village 20

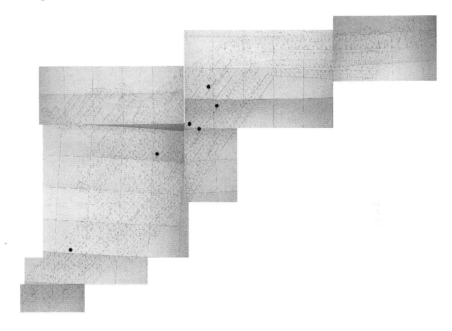

Village 21

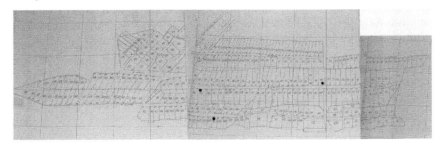

Village 22

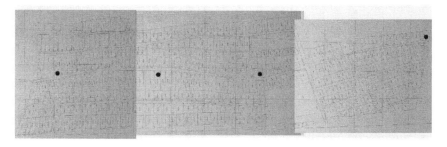

Village 23

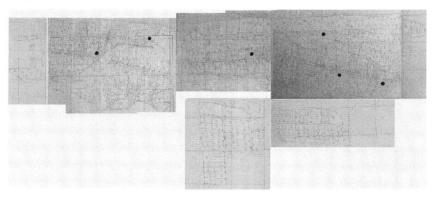

Village 24

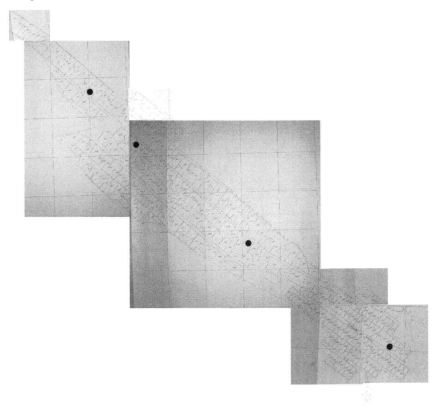

Village 25

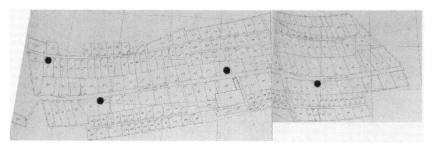

Village 26

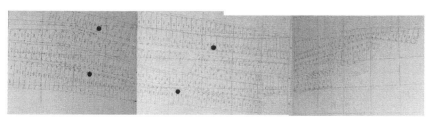

Village 27

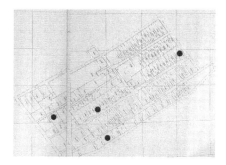

Village 28

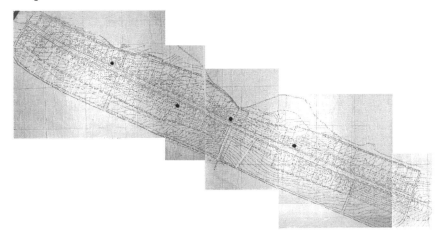

Village 29

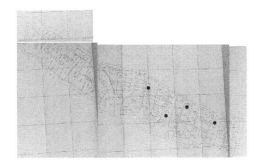

Village 30

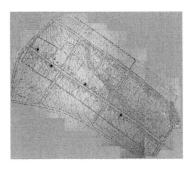

D   Additional Agricultural Statistics of the Sample

This appendix present information describing additional parameters of the sampled households' agricultural activity, which cannot be described in the main text, due to size limitation, but could still be interesting for a better understanding of the production features in the target area. The information includes more details about the cost side of the production activity; production works information, water supply, use of fertilizers, etc.

Table D.1: Share of the Rural Households, Implementing Agricultural Production Activity

		Total Oblast*	Talas	Bakai-Ata	Kara-Buura	Manas
1	Households:					
a	cultivating any crop	245	63	71	67	44
b	cultivating haricot beans	140	56	19	33	32
c	cultivating other crops	158	13	69	61	15
2	Households:					
a	breeding any livestock	212	55	52	61	44
b	breeding sheep	143	47	36	36	24
c	breeding cattle	176	46	39	53	38
d	breeding horses	42	19	10	9	4

Source: Survey Data
*Notes:
a) Total number of the households in the sample is 297.
b) There are described crop production executed only on the main land plot.

Table D.2: Average Basic Crop Production Costs, Som per Household

		Total Oblast*	Talas	Bakai-Ata	Kara-Buura	Manas
1	Mechanized works	8891	9626	7624	10123	8007
2	Hired laborers	4754	5804	4650	5175	2843
3	Production input costs					
a	haricot beans	13974	19336	12077	12344	24671
b	other crops	2094	1922	2011	3454	1040
4	Irrigation	870	437	699	1247	1190
5	Taxes	1359	780	1217	1933	1543

Source: Survey Data
*Notes:
a) Main costs are described in details in the tables below, for other small costs explanation is provided.
b) Hired laborers cost includes costs of manual works for harvesting of the haricot beans, and in some cases for the potatoes; from the sample of 297 households, 122 households is reported about this type of costs, costs are given in Som per Household.
c) Irrigation costs are given in Som per hectare per Household.
d) Taxes includes Land Tax and Social Tax, costs are given in Som per Household.

Table D.3: Mechanized Works Average Costs, Som per Household

		Total Oblast*	Talas	Rayons Bakai-Ata	Kara-Buura	Manas
1	Leveling	633	337	697	736	798
2	Ploughing	3528	3261	3576	4188	2831
3	Seeding	1676	1214	1692	2080	1693
4	Harvesting	1810	3555	464	1563	1860
5	Other works	1243	1259	1194	1556	825
	Total mechanical works	8891	9626	7624	10123	8007

Source: Survey Data
*Notes:
a) Other works includes mainly costs of the mechanical weed cleaning and mowing of the hay.

Table D.4: Production Input Costs for Haricot Beans Producers, Som per Household

		Total Oblast*	Talas	Rayons Bakai-Ata	Kara-Buura	Manas
1	Seeds	1335	1631	1085	814	4347
2	Mineral Fertilizers	4736	4700	4072	4054	10594
3	Organic fertilizers	77	0	133	8	163
4	Herbicides, Pesticides	235	727	44	275	523
5	Packaging	688	2598	471	541	633
6	Transportation	1617	3038	1016	1300	4433
7	Storage	22	0	22	0	133
8	Harvesting costs	5264	6642	5235	5351	3844
	Total Costs	13974	19336	12077	12344	24671

Source: Survey Data
*Notes:
Description is given in the table D.5.

Table D.5: Additional Harvesting Costs for Households Cultivating Haricot Beans, Som per Household

		Total		Rayons		
		Oblast*	Talas	Bakai-Ata	Kara-Buura	Manas
1	Costs for clearing seeds					
	haricot beans, kg	148	191	148	149	111
	fuel, liters	30	38	30	30	22
2	Costs for clearing seeds, Som					
	haricot beans	4372	5496	4343	4459	3176
	fuel	892	1146	892	893	668
	Total Cost	5264	6642	5235	5351	3844

Source: Survey Data
*Notes:
Haricot beans harvested manually, but clearing of the beans from pods is mechanized; this service costs consist of two components: costs of fuel (10 liters for 1 ton of processed beans), and clearing costs (50 kg for 1 ton of processed beans). Service is provided in the fields at harvest period. Amount of beans paid for clearing deducted from the harvest estimation.

Table D.6: Production Input Costs for Other Crops, Som per Household

		Total		Rayons		
		Oblast	Talas	Bakai-Ata	Kara-Buura	Manas
1	Seeds	371	457	53	345	438
2	Mineral Fertilizers	48	42	105	48	22
3	Organic fertilizers	41	95	0	15	0
4	Herbicides, Pesticides	44	57	6	88	0
5	Packaging	266	318	389	340	26
6	Transportation	1274	954	1458	2617	341
7	Storage	49	0	0	0	214
	Cost for production	2094	1922	2011	3454	1040

Source: Survey Data

Table D.7: Livestock Production Costs, Som per Household

		Total Oblast*	Talas	Bakai-Ata	Kara-Buura	Manas
1	Veterinary services					
a	Sheep	604	446	600	928	433
b	Cattle	176	293	171	156	100
c	Horses	120	127	99	133	105
2	Pasture grazing costs					
a	Sheep	6578	5612	6914	10075	2721
b	Cattle	2636	3302	2731	2699	1644
c	Horses	7032	5798	4760	13778	3400
3	Purchased forage					
a	grain forage	2198	3019	1864	2809	719
b	hay	5659	5100	9356	5643	2010
c	other forage	409	248	415	314	736

Source: Survey Data
*Notes:
a) Veterinary services includes costs of prevention vaccination of livestock from typical diseases.
b) Pasture grazing costs includes costs of grazing in near-village pastures (winter pastures) and remote pastures (summer pastures); rates are based on the price per head of livestock.
c) Forage is purchased additionally for winter period and do not include own produced or collected forage.

E TSLS First Stage Results

This appendix presents the results of the first stage of the two-stage least squares regression of the poverty-efficiency linkage. The final results of this model are presented in Table 6.9 on page 128. The table below shows the LS estimation coefficients for the land quality (LQ) and land distance(DL) variables as explanatory variables and the technical efficiency estimates as the dependant variable. A separate correlation is shown for each explanatory variable set, as well as for the whole set. The values for the those households, which do not have their own land or do not know the answer, are out of the first stage, in the purpose to prevent a perfect multicollinearity case. Thus, in each set of variables, estimations are implemented towards those households, which do not have their own land or do not know the answer. Explanatory variables (instruments) description is given on page 127.

Table E.1: First stage of Two-Stage Least Squares Estimation of SFA efficiency and Instrumental Variables

Dep. Var.	Coef.		SFA*	
No. obs.			289	
Const.	C(1)	0.47	0.49	0.47
		(0.03)	(0.02)	(0.02)
Distance to Land Plot				
$DL1$	C(2)	0.14	-	0.15
		(0.06)		(0.03)
$DL2$	C(3)	0.13	-	0.14
		(0.06)		(0.03)
$DL3$	C(4)	0.13	-	0.13
		(0.06)		(0.03)
$DL4$	C(5)	0.1	-	0.1
		(0.07)		(0.05)
Land Quality				
$LQ1$	C(6)	-0.06	0.05	-
		(0.06)	(0.03)	
$LQ2$	C(7)	0.001	0.11	-
		(0.06)	(0.03)	
$LQ3$	C(8)	0.04	0.16	-
		(0.06)	(0.03)	
$LQ4$	C(9)	-0.02	0.09	-
		(0.16)	(0.15)	
R^2		0.11	0.09	0.07
F-statistic		4.4	7.5	5.8

Source: LS Estimations Results from Survey Data
*Notes:
a) standard errors are given in parentheses.

Table E.2: First stage of Two-Stage Least Squares Estimation of DEA CRS efficiency and Instrumental Variables

Dep. Var.	Coef.	$DEA1_{CRS}^*$		
No. obs.		289		
Const.	C(1)	0.18	0.20	0.18
		(0.04)	(0.03)	(0.04)
Distance to Land Plot				
$DL1$	C(2)	0.15	-	0.20
		(0.08)		(0.05)
$DL2$	C(3)	0.16	-	0.21
		(0.08)		(0.04)
$DL3$	C(4)	0.16	-	0.20
		(0.08)		(0.05)
$DL4$	C(5)	0.03	-	0.06
		(0.10)		(0.07)
Land Quality				
$LQ1$	C(6)	-0.04	0.08	-
		(0.08)	(0.05)	
$LQ2$	C(7)	0.04	0.17	-
		(0.08)	(0.04)	
$LQ3$	C(8)	0.12	0.25	-
		(0.08)	(0.04)	
$LQ4$	C(9)	-0.14	-0.01	-
		(0.22)	(0.21)	
R^2		0.15	0.12	0.09
F-statistic		6.2	10.4	7.4

Source: LS Estimations Results from Survey Data
*Notes:
a) standard errors are given in parentheses.

Table E.3: First stage of Two-Stage Least Squares Estimation of DEA Labor Scale Efficiency and Instrumental Variables

Dep. Var.	Coef.	$DEA2_{SCLAB}{}^*$		
No. obs.		289		
Const.	C(1)	0.04	0.05	0.05
		(0.03)	(0.02)	(0.02)
Distance to Land Plot				
$DL1$	C(2)	0.07	-	0.18
		(0.06)		(0.03)
$DL2$	C(3)	0.11	-	0.22
		(0.06)		(0.03)
$DL3$	C(4)	0.08	-	0.18
		(0.06)		(0.03)
$DL4$	C(5)	-0.01	-	0.09
		(0.07)		(0.05)
Land Quality				
$LQ1$	C(6)	0.07	0.14	-
		(0.06)	(0.03)	
$LQ2$	C(7)	0.11	0.19	-
		(0.06)	(0.03)	
$LQ3$	C(8)	0.16	0.24	-
		(0.06)	(0.03)	
$LQ4$	C(9)	-0.05	0.01	-
		(0.16)	(0.16)	
R^2		0.20	0.17	0.15
F-statistic		8.8	14.8	13.1

Source: LS Estimations Results from Survey Data
*Notes:
a) standard errors are given in parentheses.

Bibliography

Abazov, R. (1999a). Economic Migration in Post-Soviet Central Asia: The Case of Kyrgyzstan. *Post-Communist Economies, 11*(2), 237–252.

Abazov, R. (1999b). Policy of Economic Transition: in Kyrgyzstan. *Central Asian Survey, 18*(2), 197–223.

Abazov, R. (2004). *Historical Dictionary of Kyrgyzstan*, Volume 49. Scarecrow Press, Inc., Maryland, USA.

Abazov, R. (2008). *The Palgrave Concise Historical Atlas of Central Asia*. Number ISBN 1-4039-7542-6. Palgrave Macmillan, New York, USA.

Abdul Wahab, M. (1980). Income and Expenditure Surveys in Developing Countries: Sample Design and Execution. Technical report, World Bank, Development Research Center, LSMS, Washington, D.C., USA.

Ackland, R. (1996). The 1993 Kyrgyz Multipurpose Poverty Survey (KPMS): Documentation. Technical report, World Bank, Poverty and Human Resource Division, Policy and Research Department, Washington, D.C., USA.

Adle, C., I. Habib, and K. M. Baipakov (Eds.) (2003). *History of Civilizations of Central Asia. Development in Contrast: From the Sixteenth to the Mid-Nineteenth Century*, Volume V of *Multiple History Series*. UNESCO Publishing, Paris, France.

Adle, C., M. K. Palat, and A. Tabyshalieva (Eds.) (2005). *History of Civilizations of Central Asia. Towards the Contemporary Period: From the Mid-Nineteenth to the End of the Twentieth Century*, Volume VI of *Multiple History Series*. UNESCO Publishing, Paris, France.

Afrait, S. (1972). Efficiency Estimation of Production Functions. *International Economic Review, 13*(3), 568–598.

Aigner, D. J. and S. Chu (1968). On Estimating the Industry Production Function. *American Economic Review, 58*, 826–839.

Aigner, D. J., K. C. Lovell, and P. Schmidt (1976). Formulation and Estimation of Stochastic Frontier Production Function Model. *Journal of Econometrics, 6*, 21–37.

Ainsworth, M. and J. van der Gaag (1988). Guidelines For Adapting the LSMS Living Standards Questionnaires to Local Conditions. Technical report, World Bank, Development Research Center, LSMS, Washington, D.C., USA.

Aslund, A. (2007). *How Capitalism Was Built. The Transformation of Central and Eastern Europe, Russia, and Central Asia*. Number ISBN 0-521-68382-3. Cambridge University Press, New York, USA.

Banker, R., A. Charnes, and W. Cooper (1984). Some Models for Estimating Technical and Scale Inefficiencies in Data Envelopment Analysis. *Management Science, 30*, 1078–1092.

Battese, G. E. and T. J. Coelli (1988). Prediction of Firm-Level Technical Efficiencies With a Generalised Frontier Production Function and Panel Data. *Journal of Econometrics, 38*, 387–399.

Battese, G. E. and T. J. Coelli (1992). Frontier Production Functions, Technical Efficiency and Panel Data: With Application to Paddy Farmers in India. *Journal of Productivity Analysis, 3*, 153–169.

Battese, G. E. and G. Corra (1977). Estimation of a Production Frontier Model: With Application to the Pastoral Zone of Eastern Australia. *Australian Journal of Agricultural Economics,* (21), 169–179.

Bravo-Ureta, B. E. and A. E. Pinheiro (1993). Efficiency Analysis of Developing Country Agriculture: A Review of the Frontier Function Literature. *American Economic Review,,* 88–101.

Bregel, Y. (2003). *An Historical Atlas of Central Asia*. Number ISBN 90 04 12321 0. Brill Academic Publishers, Leiden, Netherlands.

Chander, R., C. Grootaert, and G. Pyatt (1980). Living Standards Surveys in Developing Countries. Technical report, World Bank, Development Research Center, LSMS, Washington, D.C., USA.

Charnes, A., W. Cooper, and E. Rhodes (1978). Measuring the Efficiency of Decision Making Units. *European Journal of Operations Research, 2*, 429–444.

Coelli, T. J. (1995). Estimators and Hypothesis Tests for a Stochastic: A Monte Carlo Analysis. *Journal of Productivity Analysis,* (6), 247–268.

Coelli, T. J. (1996a). *A Guide to DEAP Version 2.1: A Data Envelopment Analysis (Computer) Program*. Centre for Efficiency and Productivity Analysis, University of New England, Australia, http://www.uq.edu.au/economics/cepa/deap.php.

Coelli, T. J. (1996b). *A Guide to FRONTIER Version 4.1: A Computer Program for Stochastic Frontier Production and Cost Function Estimation.* Centre for Efficiency and Productivity Analysis, University of New England, Australia, http://www.uq.edu.au/economics/cepa/frontier.php.

Coelli, T. J., P. D. Rao, C. J. O'Donnell, and G. E. Battese (2005). *An Introduction to Efficiency and Productivity Analysis* (2 ed.). Number ISBN 0-387-24265-1. Springer, USA.

Daraio, C. and L. Simar (2007). *Advanced Robust and Nonparametric Methods in Efficiency Analysis.* Number ISBN 0-387-35155-8. Springer, USA.

Deaton, A. (1997). *The Analysis of Household Surveys: A Microecnometric Approach to Development Policy.* The John Hopkins University Press, Maryland, USA.

Deaton, A. and S. Zaidi (2002). Guideliness for Construction Consumption Aggregates for Welfare Analysis. Technical report, World Bank, Development Research Center, LSMS, Washington, D.C., USA.

Debreu, G. (1951). The Coefficient of Resource Utilization. *Econometrica, 19*(3), 273–292.

Esenaliev, D., A. Kroeger, and S. Steiner (2011). The Kyrgyz Integrated Household Survey (KIHS): A Primer. Technical Report 62, German Institute for Economic Research (DIW-Berlin), Germany.

Everett-Heath, T. (Ed.) (2003). *Central Asia: Aspects of Transition.* Number ISBN 0 700 70957 6. Central Asia Research Forum, Routledge Curzon, London, UK.

Färe, R. and S. Grosskopf (2004). *New Directions: Efficiency and Productivity.* Number ISBN 1-4020-7661-4. Kluwer Academic Publishers, USA.

Färe, R., S. Grosskopf, and K. C. Lovell (1994). *Production Frontiers.* Number ISBN 0 521 42033 4. Cambridge University Press, Cambridge, UK.

Farrell, M. (1957). The Measurement of the Productive Efficiency. *Journal of the Royal Statistical Society, 120*(3), 253–290.

Fields, G. (1994). Data for Measuring Poverty and Inequality Changes in the Developing Countries. *Journal of Development Economics, 44*(1), 81–102.

Forsund, F. R., K. C. Lovell, and P. Schmidt (1980). Survey of Frontier Production Functions and Their Relationships to Efficiency Measurement. *Journal of Econometrics, 13*, 5–25.

Foster, J. (1984). On Economic Poverty: A Survey of Aggregate Measures. In R. Basman and G. Rhodes (Eds.), *Advances in Econometrics*, Chapter 3. JAI Press, Connecticut, USA.

Foster, J., J. Greer, and E. Thorbecke (1984). A Class of Decomposable Poverty Measures. *Econometrica, 52*(3), 761–766.

Greene, W. H. (1980). Maximum Likelihood Estimation of Econometric Frontier Functions. *Journal of Econometrics, 13*(1), 27–56.

Grosh, M. E. and J. Muñoz (1996). A Manual for Planning and Implementing the Living Standards Measurent Study Survey. Technical report, World Bank, Development Research Center, LSMS, Washington, D.C., USA.

Haughton, J. (2005). Introduction to Poverty Analysis. Technical report, World Bank Institute, Washington, D.C., USA.

Heston, A., R. Summers, and B. Aten (2012). Penn World Table. Version 7.0. Technical report, Center for International Comparisons of Production, Income and Prices at the University of Pennsylvania, USA.

Himmelblau, D. M. (1972). *Applied Non-Linear Programming.* McGraw-Hill, New York, USA.

Howes, S. and J. O. Lanjouw (1997). Poverty Comparisons and Household Survey Design. Technical report, World Bank, Development Research Center, LSMS, Washington, D.C., USA.

Ibragimova, S. (2008). Country Report on International Migrants' Remittances and Poverty in the Kyrgyz Republic. Technical report, Asian Development Bank, Bishkek, Kyrgyzstan.

ICG (2005). Kyrgyzstan: After Revolution. Technical Report 97, International Crisis Group, http://www.crisisgroup.org/en/regions/asia/central-asia/kyrgyzstan/097-kyrgyzstan-after-the-revolution.aspx.

KIC (2011). Report of the Independent International Comission of Inquiry into the Events Southern Kyrgyzstan in June 2010. Technical report, Kyrgyzstan Inquiry Commission, International Commision on the Kyrgyz Violence in 2010, http://www.oscepa.org/parliamentary-diplomacy/special-representative-visits/614-kyrgyzstan-inquiry-commission-releases-final-report.

Kish, L. (1965). *Survey Sampling.* Number ISBN 0 4711 0949 5. John Wiley and Sons, Inc., New York, USA.

KMPS (1994a). Kyrgyz Multipurpose Poverty Survey- Survey of availabilty and prices of food products and fuel. Technical report, World Bank, KMPS, Washington, D.C., USA.

KMPS (1994b). Kyrgyz Multipurpose Poverty Survey. Household Questionnaire. Technical report, World Bank, KMPS, Washington, D.C., USA.

KMPS (1994c). Kyrgyz Multipurpose Poverty Survey. Questionnaire for Adults. Technical report, World Bank, KMPS, Washington, D.C., USA.

KMPS (1994d). Kyrgyz Multipurpose Poverty Survey. Questionnaire for Children. Technical report, World Bank, KMPS, Washington, D.C., USA.

KMPS (1994e). Kyrgyz Multipurpose Poverty Survey. Questionnaire for Communities. Technical report, World Bank, KMPS, Washington, D.C., USA.

Koopmans, T. (1951). An Analysis of Production as an Efficient Combination of Activities. In T. Koopmans (Ed.), Activity Analysis of Production and Allocation. Wiley, New York, USA.

Kort, M. (2004). *Nations in Transition: Central Asian Republics.* Number ISBN 0-8160-5074-0. Facts On File Inc., New York, USA.

KPMS (2002a). Kyrgyz Poverty Monitoring Survey - Fall 1998 Household Questionnaire. Technical report, World Bank, KPMS, Washington, D.C., USA.

KPMS (2002b). Kyrgyz Poverty Monitoring Survey - Fall 1998 Population Point Questionnaire. Technical report, World Bank, Washington, D.C., USA.

Kumbhakar, S. C. and K. C. Lovell (2000). *Stochastic Frontier Analysis.* Number ISBN 0-521-85761-9. Cambridge University Press, Cambridge, UK.

Lau, L. and P. Yotopordos (1971). A Test for Relative Efficiency and Applications to Indian Agriculture. *American Economic Review, 61,* 94–109.

Lee, L. and W. Tyler (1978). The Stochastic Frontier Production Function and Average Efficiency. *Journal of Econometrics, 7*(3), 385–389.

Lehrman, Z. and D. Sedik (2009). Agrarian Reform in Kyrgyzstan: Achievements and the Unfinished Agenda. Technical Report 1, Policy Studies on Rural Transition, FAO Regional Office for Europe and Central Asia, http://www.fao.org.

Martino, L. D., M. Vahobova, and D. Gullette (2009). Central Asia Risk Assessment: Responding to Water, Energy, and Food Insecurity. Technical report, Regional Bureau for Europe and the CIS, United Nations Development Programme, New York, USA.

Matthews, M. (1986). *Poverty in the Soviet Union.* Number ISBN 0-521-32544-8. Cambridge University Press, New York, USA.

McMann, K. M. (2006). *Economic Autonomy and Democracy: Hybrid Regimes in Russia and Kyrgyzstan.* Number ISBN 0-521-85761-9. Cambridge University Press, New York, USA.

Meeusen, W. and J. van den Broeck (1977). Efficiency Estimation from Cobb-Douglas Production Functions with Composed Error. *International Economic Review, 18*(2), 435–444.

NSC (2008). Kyrgyz Integrated Household Survey Questionnaire (Russian Version). Technical report, National Statistical Committee, Bishkek, Kyrgyzstan.

NSC (2010a). National Census of the Kyrgyz Republic 2009. Volume 3. Talas Oblast (Russian Version). Technical report, National Statistical Committee, Bishkek, Kyrgyzstan.

NSC (2010b). Poverty Indicators of Kyrgyz Republic in 2008 (Russian Version). Technical report, National Statistical Committee, Bishkek, Kyrgyzstan, http://www.stat.kg.

NSC (2011a). Demographic Yearbook of the Kyrgyz Republic 2006-2010 (Russian Version). Technical report, National Statistical Committee, Bishkek, Kyrgyzstan, http://www.stat.kg.

NSC (2011b). Index of Consumer Prices and Tariffs in Kyrgyz Republic for June, 2011 (Russian Version). Technical report, National Statistical Committee, Bishkek, Kyrgyzstan, http://www.stat.kg.

NSC (2011c). Living Standards of Population of the Kyrgyz Republic 2006-2010 (Russian Version). Technical report, National Statistical Committee, Bishkek, Kyrgyzstan, http://www.stat.kg.

NSC (2012a). Average Market Prices for Food Products in Kyrgyz Republic 2010-2011 (Russian Version). Technical report, National Statistical Committee, Bishkek, Kyrgyzstan, http://www.stat.kg.

NSC (2012b). Exchange Rates of Basic Currencies in Kyrgyz Republic in 2007-2011 (Russian Version). Technical report, National Statistical Committee, Bishkek, Kyrgyzstan, http://www.stat.kg.

NSC (2012c). Poverty Analysis Methods (Russian Version). Technical report, National Statistical Committee, Bishkek, Kyrgyzstan, http://www.stat.kg.

OECD (2005). Fighting Corruption in Transition Economies: Kyrgyz Republic. Technical Report ISBN 92-64-01083-1, Organisation for Economic Co-Operation and Development, Paris, France.

OSCE (2012). Kyrgyz Republic Presidential Election 30 October 2011 OSCE ODIHR Election Observation Mission Final Report. Technical report, Office for Democratic Institutions and Human Rights, Organisation for Security and Co-Operation in Europe, Warsaw, Poland.

Pitt, M. M. and L.-F. Lee (1981). The Measurement and Sources of Technical Inefficiency in the Indonesian Weaving Industry. *Journal of Development Economics, 9*(1), 43–64.

Pomfret, R. and K. H. Anderson (1999). Poverty in the Kyrgyz Republic. *Asia-Pacific Development Journal, 6*(1), 73–91.

Ravallion, M. (1992). Poverty Comparisons. A Guide to Concepts and Methods. Technical report, World Bank, Development Research Center, LSMS, Washington, D.C., USA.

Ravallion, M. (1998). Poverty Lines in Theory and Practice. Technical report, World Bank, Development Research Center, LSMS, Washington, D.C., USA.

Richmond, J. (1974). Estimating the efficiency of production. *International Economic Review, 15*(2), 515–521.

Ritter, C. and L. Simar (1997). Pitfalls of Normal-Gamma Stochastic Frontier Models. *Journal of Productivity Analysis, 8*(2), 167–182.

Sen, A. (1976). Poverty: An Ordinal Aapproach to Measurement. *Econometrica, 44*(2), 219–231.

Temesgen, T., S. Ibragimova, and D. Steele (2002). Kyrgyz Poverty Monitoring Surveys (KPMS), Fall 1996 - Fall 1998, Basic Information Document. Technical report, World Bank, KPMS, Washington, D.C., USA.

Tilekeyev, K. (2010). Dynamics of Poverty in Transition in Kyrgyzstan. Master's thesis, Justus-Liebeig University Giessen (unpublished), Giessen, Germany.

TSO (2011). Statistical Yearbook 2010. Talas Oblast. Technical report, Talas Statistical Office, National Statistical Committee, Talas, Kyrgyzstan.

UNDP (2005). Central Asia Human Development Report Bringing Down Barriers: Regional Cooperation for Human Development and Human Security. Technical Report ISBN 92-95042-34-4, United Nation Development Program, Regional Bureau for Europe and the CIS, Bratislava, Slovak Republic.

UNDP (2006a). The Influence of Civil Society on the Human Development Process in Kyrgyzstan. Technical Report ISBN 9967-23-706-6, United Nation Development Program, Bishkek, Kyrgyzstan.

UNDP (2006b). The Shadow Economy in the Kyrgyz Republic: Trends, Estimates and Policy Options. Technical report, United Nations Development Programme, Bishkek, Kyrgyzstan.

Watts, H. W. (1968). An Economic Definition of Poverty. In D. Moynihan (Ed.), *On Understanding Poverty*, Chapter 1, pp. 1–21. Basic Books, New York, USA.

WB (1995). The Kyrgyz Republic. Poverty Assessment and Strategy. Technical Report 14380-KG, World Bank, Human Resource Division, Europe and Central Asia Department III, Washington, D.C., USA.

WB (1999). Kyrgyz Republic Private Sector Review in the Transitional Era. Technical Report 18121-KG, World Bank, Europe and Central Asia Region, Washington, D.C., USA.

WB (2001). Kyrgyz Republic. Poverty in the 1990s in the Kyrgyz Republic. Technical Report 21721-KG, World Bank, Human Development Department, Country Department VIII, Europe and Central Asia Region, Washington, D.C., USA.

WB (2005). Kyrgyz Republic. Poverty Update Profile of Living Standards in 2003. Technical Report 36602-KG, World Bank, Europe and Central Asia Region, Poverty Reduction and Economic Management Unit, Washington, D.C., USA.

WB (2007a). Kyrgyz Republic. Poverty Assessment Volume 1: Growth, Employment and Poverty. Technical Report 40864-KG, World Bank, Europe and Central Asia Region, Poverty Reduction and Economic Management Unit, Washington, D.C., USA.

WB (2007b). Kyrgyz Republic. Poverty Assessment Volume 2: Labour Market Dimensions of Poverty. Technical Report 40864-KG, World Bank, Europe and Central Asia Region, Poverty Reduction and Economic Management Unit, Washington, D.C., USA.

Zheng, B. (1993). An Axiomatic Characterization of the Watts Poverty Index. *Economics Letters, 42*(1), 81–86.

Zheng, B. (1997). Aggregate Poverty Measures. *Journal of Economic Surveys, 11*(2), 123–162.

Ziliak, J. P. (2005). Understanding Poverty Rates and Gaps: Concepts, Trends and Challenges. *Foundations and Trends in Microeconomics, 1*(3), 127–199.

Schriften zur Internationalen Entwicklungs- und Umweltforschung

Herausgegeben vom

Zentrum für internationale
Entwicklungs- und
Umweltforschung
der Justus-Liebig-Universität Gießen

Band   1 Hans-Rimbert Hemmer / Rainer Wilhelm: Fighting Poverty in Developing Countries. Principles for Economic Policy. 2000.

Band   2 Lorenz King / Martin Metzler / Tong Jiang (eds.): Flood Risks and Land Use Conflicts in the Yangtze Catchment, China and at the Rhine River, Germany. 2001.

Band   3 Ingrid-Ute Leonhäuser (ed.): Women in the Context of International Development and Cooperation. Review and Perspectives. Selected Papers and Abstracts presented at the Justus-Liebig-University Gießen 26.-28. October 2000. 2002

Band   4 Margit Schratzenstaller: Internationale Mobilität von und internationaler fiskalischer Wettbewerb um Direktinvestitionen. 2002.

Band   5 Armin Bohnet u.a.: Theoretische Grundlagen und praktische Gestaltungsmöglichkeiten eines Finanzausgleichssystems für die VR China. Unter Mitwirkung von Chen Biyan, Chen Shixin, Ge Licheng, Ge Naixu, Ge Zhuying, Ma Shuanyou, Markus Peplau, Yang Zhigang, Zhu Qiuxia. 2003.

Band   6 Armin Bohnet / Matthias Höher (eds.): The Role of Minorities in the Development Process. 2004.

Band   7 Thi Phuong Hoa Nguyen: Foreign Direct Investment and its Contributions to Economic Growth and Poverty Reduction in Vietnam (1986–2001). 2004.

Band   8 Andreas Böcker / Roland Herrmann / Michael Gast / Jana Seidemann: Qualität von Nahrungsmitteln. Grundkonzepte, Kriterien, Handlungsmöglichkeiten. 2004.

Band   9 Christina Mönnich: Tariff Rate Quotas and Their Administration. Theory, Practice and an Econometric Model for the EU. 2004.

Band 10 Reimund Seidelmann / Ernst Giese (eds.): Cooperation and Conflict Management in Central Asia. 2004.

Band 11 Claudia Ohly: Das Steuersystem im ungarischen Transformationsprozess. Ein Beitrag zur Transformationstheorie. 2004.

Band 12 Nicole Mau: Umweltzertifikate. Der Einsatz von Umweltzertifikaten in der Landwirtschaft am Beispiel klimarelevanter Gase. 2005.

Band 13 P. Michael Schmitz (Hrsg.): Water and Sustainable Development. 2005.

Band 14 Ira Pawlowski: Die Wettbewerbsfähigkeit der ukrainischen Milchwirtschaft. Messung von Marktverzerrung und Politikeinfluß im Transformationsprozeß. 2005.

Band 15 Kirsten Westphal (ed.): A Focus on EU-Russian Relations. Towards a close partnership on defined road maps? 2005.

Band 16 Andreas Langenohl / Kirsten Westphal (eds.): Conflicts in a Transnational World. Lessons from Nations and States in Transformation. 2006.

Band 17 Rosemarie von Schweitzer: Home Economics Science and Arts. Managing Sustainable Everyday Life. 2006.

www.peterlang.com